Aberration Theory
Made Simple

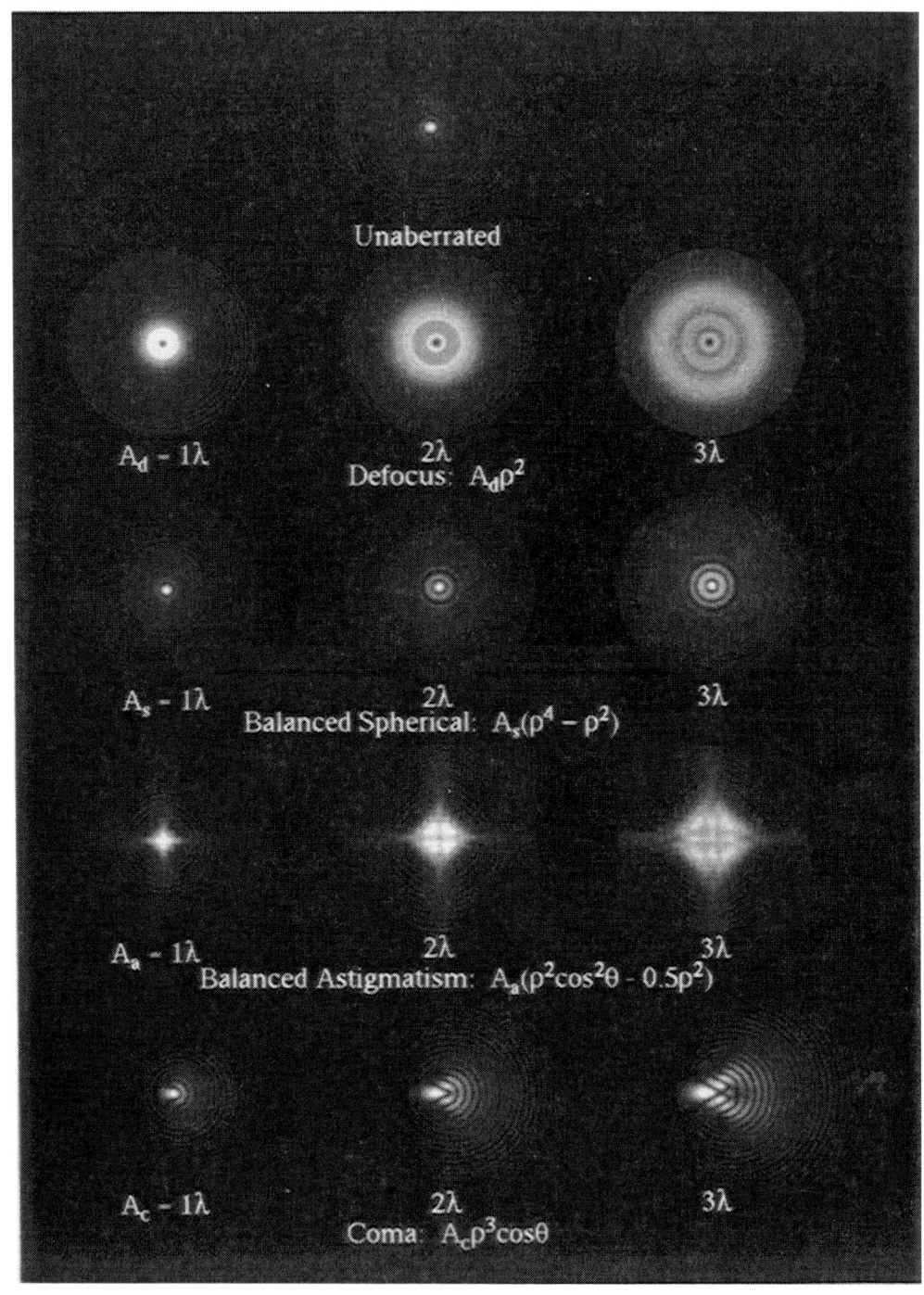

Unaberrated and aberrated images of a point object as indicated. For details, see Sec. 8.4. Photo (this page and cover) by Dr. Richard Boucher, The Aerospace Corporation.

Aberration Theory
Made Simple

Virendra N. Mahajan

The Aerospace Corporation and
University of Southern California

Donald C. O'Shea, Series Editor
Georgia Institute of Technology

TUTORIAL
TEXTS
IN OPTICAL
ENGINEERING

Volume TT 6

SPIE OPTICAL ENGINEERING PRESS

A Publication of SPIE—The International Society for Optical Engineering
Bellingham, Washington USA

Library of Congress Cataloging-in-Publication Data

Mahajan, Virendra N.,
    Aberration theory made simple / Virendra N. Mahajan
        p. cm. — (Tutorial texts in optical engineering : v. TT 6)
    Includes bibliographical references and index.
    ISBN 0-8194-0536-1
    1. Aberration. 2. Imaging Systems. 3. Optics, Geometrical.
  I. Title. II. Series.
  QC671.M34 1991
  621.36—dc20   90-48633
                                               CIP

Published by
SPIE—The International Society for Optical Engineering
P.O. Box 10
Bellingham, Washington 98227-0010

Copyright © 1991 The Society of Photo-Optical Instrumentation Engineers

Printed in the United States of America

Fourth Printing

To My

Wife	**Shashi Prabha**
Son	**Vinit Bharati**
Daughter	**Sangita Bharati**

Books in the SPIE Tutorial Texts Series

Introduction to the Series

These Tutorial Texts provide an introduction to specific optical technologies for both professionals and students. Based on selected SPIE short courses, they are intended to be accessible to readers with a basic physics or engineering background. Each text presents the fundamental theory to build a basic understanding as well as the information necessary to give the reader practical working knowledge. The included references form an essential part of each text for the reader requiring a more in-depth study.

Many of the books in the series will be aimed to readers looking for a concise tutorial introduction to new technical fields, such as CCDs, fiber optic amplifiers, sensor fusion, computer vision, or neural networks, where there may be only limited introductory material. Still others will present topics in classical optics tailored to the interests of a specific audience such as mechanical or electrical engineers. In this respect the Tutorial Text serves the function of a textbook. With its focus on a specialized or advanced topic, the Tutorial Text may also serve as a monograph, although with a marked emphasis on fundamentals.

As the series develops, a broad spectrum of technical fields will be represented. One advantage of this series and a major factor in the planning of future titles is our ability to cover new fields as they are developing, giving people the basic knowledge necessary to understand and apply new technologies.

Donald C. O'Shea   June 1991
Georgia Institute of Technology

Foreword

It is a distinct pleasure for me to write this short foreword to Dr. Virendra Mahajan's tutorial text, *Aberration Theory Made Simple*. I write it not because I am particularly knowledgeable about aberration theory—in fact, it may be because I am not particularly knowledgeable that I was invited! This is a tutorial text, and as a lifelong educator I am also a lifelong learner and I should be able to learn from this text; and I did.

This text is prepared in the ideal way for a tutorial. It comes as a direct result of teaching this material to a wide range of audiences in a wide range of locations; so it has been tried and tested. The "student guinea pigs" have performed their invaluable service so that those of us who come along later have the benefit of their and the author's labors.

Dr. Mahajan has lived up to his title and made aberration theory simple. Of course, I should caution the reader that simple is relative. Some topics do not yield easily to simple yet accurate descriptions. Those readers who insist that "rays" are the most important components of any analysis of optical systems, whether aberrant or not, will be very satisfied with the first half of the book, but may wish to ignore the second half. They should not. Those who are enamored with the wave approach (like me) will immediately read the second half of this book and applaud, but not go back and read the first half. They should! I did!

I am pleased that Dr. Mahajan has provided a significant list of references in addition to the bibliography at the end of the book. This will be of considerable value to the reader. Not incidentally, SPIE Optical Engineering Press will also publish a Milestone volume on Aberrations in Imaging Systems with Dr. Mahajan as the co-editor. Thus, each of us will be able to have an authoritative companion volume that contains reprints from the world's literature that will no doubt verify that this current tutorial text is indeed *Aberration Theory Made Simple*.

Brian J. Thompson
Rochester, New York

June 1991

Table of Contents

PREFACE

Aberration theory is a subject that is as old and fascinating as the field of optics. It is, however, a cumbersome subject that many students of optics do not appreciate fully. The purpose of this tutorial book is to provide a clear, concise, and consistent exposition of what aberrations are, how they arise in optical imaging systems, and how they affect the quality of images formed by them. Its emphasis is on physical insight, problem solving, and numerical results. It is intended for engineers and scientists who have a need and/or a desire for a deeper and better understanding of aberrations and their role in optical imaging and wave propagation. Although some knowledge of Gaussian optics and an appreciation for aberrations would be useful, they are not pre-requisites. What is needed is dedication and perseverance. A novice trying to learn this subject without investing much time will probably be disappointed in spite of the title of the book. The book is not intended for teaching lens design or optical testing. However, it is hoped that those working in these fields will benefit from it. It should be useful to students who may want to learn aberration theory without having to go through any lengthy derivations. These derivations are omitted out of necessity for brevity and in keeping with the spirit of these tutorials.

These tutorials have been adapted from my lectures for a graduate course entitled "Advanced Geometrical Optics," which I have been teaching in the Electrical Engineering–Electrophysics Department of the University of Southern California since 1984. They were originally developed for a short course on optical imaging and aberrations, which I taught at The Aerospace Corporation to Aerospace and Air Force personnel. They were then expanded for a short course I have been teaching at the Optical Society of America and SPIE meetings. Generally speaking, only the primary aberrations of optical systems are discussed here; they provide the first and a significant step beyond Gaussian imaging. Although a knowledge of these aberrations is very useful, they may not sufficiently describe the imaging properties of a high-quality optical system. Higher-order aberrations in such systems are often determined by ray tracing them.

This book is organized in two parts: Part I is on ray geometrical optics and Part II is on wave diffraction optics. The first chapter introduces the concepts of aperture stop and entrance and exit pupils of an optical imaging system. The wave and ray aberrations are defined and wavefront defocus and tilt aberrations are discussed. Various forms of the primary aberration function of a rotationally symmetric system are given, and how this function changes as the aperture stop of the system is moved from one position to another is discussed. The aberration function for the simplest imaging system, namely, a single spherical refracting surface, is given. Finally, a procedure by which the aberration function of a multielement system may be calculated is described. This chapter provides a foundation for the next six chapters.

Chapters 2–6 give the primary aberrations of simple systems, such as a thin lens, plane-parallel plate, spherical mirror, Schmidt camera, and a conic mirror. Numerical problems are discussed here and there to illustrate how to apply the formulas given in these chapters. Part I of the book ends with chapter 7, where the aberrated images of a point object based on geometrical optics are discussed. Thus the ray spot diagrams and, in particular, the spot sizes for primary aberrations are discussed. The concept of aberration balancing, based on geometrical optics to reduce the size of an image spot, is introduced.

In Part II, chapters 8–11 discuss the effects of aberrations on the image of a point object based on wave diffraction optics. Chapter 8 considers systems with circular exit pupils. The aberration-free characteristics of such systems are discussed in terms of their point-spread and optical transfer functions. How the aberrations affect these functions is discussed and aberration tolerances are obtained for a given Strehl or a Hopkins ratio. The concept of aberration balancing, based on wave diffraction optics, to maximize Strehl or Hopkins ratios is discussed. Systems with annular and Gaussian pupils are considered in chapter 9. The effect of obscuration on the point-spread function and on aberration tolerance is discussed. Similarly, the effect of Gaussian amplitude at the exit pupil is discussed. The content of this chapter provides a basis for assessing the effects of aberrations on the optical performance of reflecting telescopes, such as Cassegrain and Ritchey-Chrétien, and on the propagation of laser beams.

The line of sight of an aberrated system is discussed in chapter 10 in terms of the centroid of its point-spread function. It is pointed out that only coma type aberrations change the centroid. Random aberrations are considered in chapter 11, where the average point-spread and optical transfer functions for random image motion and aberrations introduced by atmospheric turbulence are discussed. Part II of the book ends with chapter 12, where a brief discussion is given on how the aberrations of a system may be observed and recognized interferometrically.

Each chapter is written to be as independent of the others as possible, although some are more so than others. For example, chapter 7 may be followed by chapter 1. Except for the first few sections of chapter 1, it is not necessary to understand Part I in order to understand Part II. However, reading Part II without Part I would be like knowing half of a story. Chapter 12 may be read at any time; however, the reason for using certain specific values of defocus, for example, in the case of spherical aberration, may not be understood unless the concepts of aberration balancing discussed in chapters 7 and 8 are understood.

On the matter of references to the literature on aberration theory, I have listed under bibliography those books that treat this subject to some or a large extent. These are the ones I have had the opportunity to read and benefit from. On the wave diffraction optics, I have given references in the text either for historical reasons (such as the papers by Airy and Lord Rayleigh) or because the work is relatively recent and has not appeared in books. Additional references are given after the bibliography for further study on part of the reader.

Finally, I would like to thank those who have helped me with the preparation of this book. I have had many discussions with Dr. Bill Swantner on geometrical optics and Dr. Richard Boucher on diffraction optics. Dr. Boucher also did computer simulations of the point-spread functions and interferograms and prepared the photographs for this book. Prof. Don O'Shea provided critical and valuable comments when he reviewed this book. Helpful comments were also provided by Prof. R. Shannon. The Sanskrit verse and its translation on p. xx were provided by Dr. S. Sutherland, University of California at Berkeley. The manuscript and its many revisions were typed by Iva Moore. The final version was produced by Betty Wenker and Candy Worshum. I thank The Aerospace Corporation for providing help and facilities to prepare this book. I also thank Dr. Roy Potter and Eric Pepper of the SPIE staff for suggesting and facilitating the preparation of this book, which was carefully edited by Rick Hermann. I cannot thank my wife and children enough for their patience during the course of this work and so I dedicate this book to them.

Los Angeles   Virendra N. Mahajan
1991

SYMBOLS AND NOTATION

a	radius of exit pupil
a_i	aberration coefficient
A_i	peak aberration coefficient
AS	aperture stop
CR	chief ray
e	eccentricity
EnP	entrance pupil
ExP	exit pupil
f	focal length
F	focal ratio or f-number, focal point
GR	general ray
h	object height
h'	image height
H	Hopkins ratio
I	irradiance
J_n	nth-order Bessel function of the first kind
L	image distance from exit pupil
m	pupil-image magnification
M	object-image magnification
MR	marginal ray
N	Fresnel number
n	refractive index
OA	optical axis
OTF	optical transfer function
p	position factor
P	object point, exit pupil power
P'	Gaussian image point
PSF	point spread function
q	shape factor
Q	aberration difference function
r_c	radius of a circle

r_0	atmospheric coherence length or diameter
R	radius of curvature of a surface or reference sphere
s	entrance pupil distance
s'	exit pupil distance
S_p	exit pupil area
S	object distance, Strehl ratio
S'	image distance
t	thickness
W	wave aberration
x,y	rectangular coordinates of a point
z	sag, observation distance
β	field angle
γ	Gaussian beam truncation parameter
δ	phase
δF	figure error
ΔR	longitudinal defocus
ϵ	obscuration ratio
r,θ	polar coordinates of a point
λ	optical wavelength
$(\xi,\eta) = \dfrac{1}{a}(x,y)$	normalized rectangular coordinates
$\rho = r/a$	normalized radial coordinate in the pupil plane
ν,ϕ	spatial frequency coordinates
σ_Φ	standard deviation of phase aberration
σ_W	standard deviation of wave aberration
τ	optical transfer function
Φ	phase aberration
ψ	angular deviation of a ray
Ψ	phase transfer function
ω	Gaussian beam radius
$R_n^m(\rho)$	radial Zernike polynomial
$<\ >$	average

अनन्तरत्नप्रभवस्य यस्य हिमं न सौभाग्यविलोपि जातम् ।
एको हि दोषो गुणसन्निपाते निमज्जतीन्दोः किरणेष्विवाङ्कः ॥

anantaratnaprabhavasya yasya himaṃ na saubhāgyavilopi jātam |
eko hi doṣo guṇasannipāte nimajjatīndoḥ kiraṇeṣv ivāṅkaḥ ||

The snow does not diminish the beauty of the Himālayan mountains which are
the source of countless gems. Indeed, one flaw is lost among a host of virtues,
as the moon's dark spot is lost among its rays.

Kālidāsa *Kumārasambhava* 1.3

PART I

Ray Geometrical Optics

CHAPTER 1

Optical Aberrations

1.1 Introduction

This chapter starts with the concepts of *aperture stop* and *entrance* and *exit pupils* of an optical imaging system. Certain special rays, such as the *chief* and the *marginal rays*, are defined. The *wave aberration* associated with a ray is defined and its relationship to the corresponding transverse *ray aberration* is given. Representations of wavefront *defocus* and *tilt* aberrations are given. We introduce different forms of the *primary aberration function* of a *rotationally symmetric system*. How this function changes as the aperture stop of the system is moved from one position to another is discussed. The primary aberration function for the simplest imaging system, namely, a single spherical refracting surface, is given for an arbitrary position of the aperture stop. Finally, we outline a procedure by which the aberration function of a multielement system may be calculated. This procedure is utilized in later chapters, for example, to calculate the aberration of a *thin lens* (Chapter 2) and a *plane-parallel plate* (Chapter 3). This chapter forms the basis of Part I on *geometrical optics*.

1.2 Optical Imaging

An optical imaging system consists of a series of refracting and/or reflecting surfaces. The surfaces refract or reflect light rays from an object to form its image. The image obtained according to geometrical optics in the *Gaussian approximation*, i.e., according to Snell's law in which the sines of the angles are replaced by the angles, is called the *Gaussian image*. The Gaussian approximation and the Gaussian image are often referred to as the *paraxial approximation* and the *paraxial image*, respectively. We assume that the surfaces are rotationally symmetric about a common axis called the optical axis (*OA*). Figures 1-1 illustrate imaging of an on-axis point object P_0 and an off-axis point object P, respectively, by an optical system consisting of two thin lenses. (For definition of a thin lens, see Section 2.2.) P' and P_0' are the corresponding Gaussian image points. An object and its image are called *conjugates* of each other, i.e., if one of the two conjugates is an object, the other is its image.

An aperture in the system which physically limits the solid angle of the rays from a point object the most is called the *aperture stop* (*AS*). For an extended (i.e., a nonpoint) object, it is customary to consider the aperture stop as the limiting aperture for the axial point object, and to determine vignetting, or blocking of some rays, by this stop for off-axis object points. The object is assumed to be placed to the left of the system so that initially light travels from left to right. The image of the stop by surfaces that precede it in the sense of light propagation, i.e., by surfaces that lie between it and the object, is called the *entrance pupil* (*EnP*). When observed from the object side, the entrance pupil appears to limit the rays entering the system to form the image of the object. Similarly, the image of the aperture stop by surfaces that follow it, i.e., by sur-

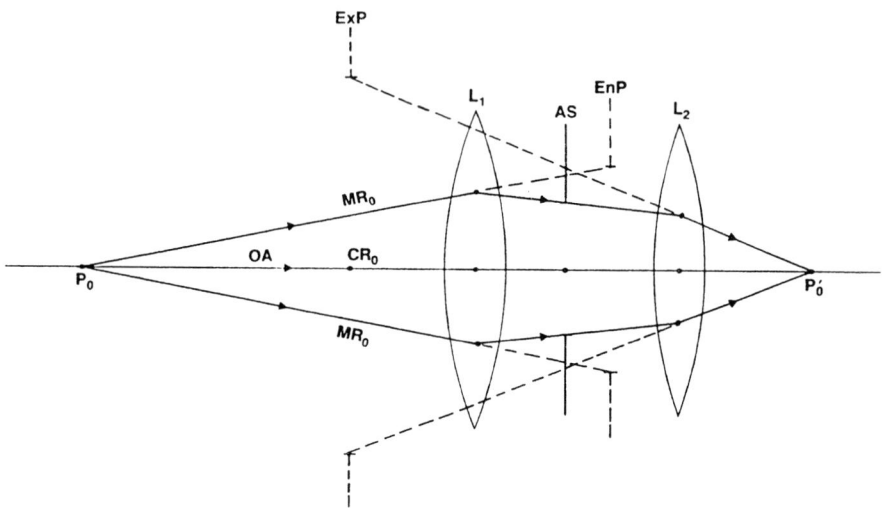

Figure 1–1a. Imaging of an on-axis point object P_0 by an optical imaging system consisting of two thin lenses L_1 and L_2. OA—optical axis, AS—aperture stop, EnP—entrance pupil, ExP—exit pupil, CR_0—chief ray, MR_0—marginal ray. P_0' is the Gaussian image of P_0.

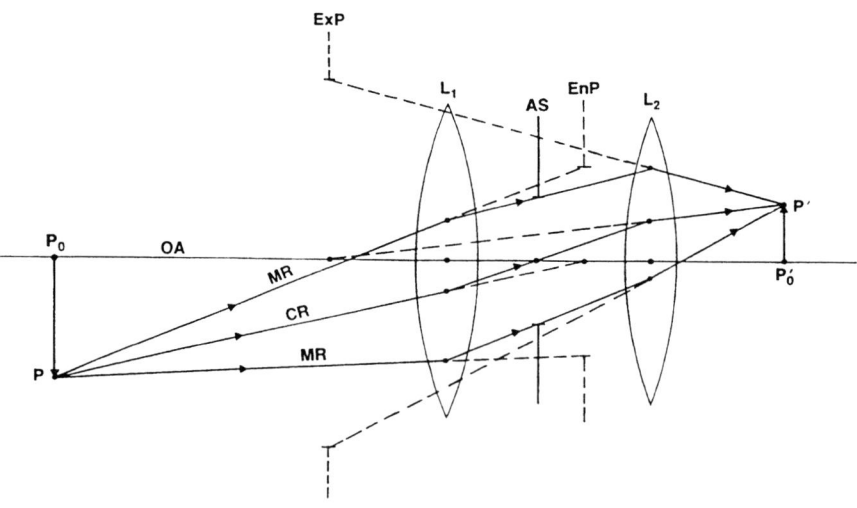

Figure 1–1b. Imaging of an off-axis point object P by an optical imaging system consisting of two thin lenses L_1 and L_2. OA—optical axis, AS—aperture stop, EnP—entrance pupil, ExP—exit pupil, CR—chief ray, MR—marginal ray. P' is the Gaussian image of P.

faces that lie between it and the image, is called the *exit pupil* (*ExP*). The object rays reaching its image appear to be limited by the exit pupil. Since the entrance and exit pupils are images of the stop by the surfaces that precede and follow it, respectively, the two pupils are conjugates of each other for the whole system; i.e., if one pupil is considered as the object, the other is its image formed by the system.

An object ray passing through the center of the aperture stop and appearing to pass through the centers of the entrance and exit pupils is called the *chief* (or the *principal*) *ray* (*CR*). An object ray passing through the edge of the aperture stop is called a *marginal ray* (*MR*). The rays lying between the center and the edge of the aperture, and, therefore, appearing to lie between the center and edge of the entrance and exit pupils, are called *zonal rays*.

It is possible that the stop of a system may also be its entrance and/or exit pupil. For example, a stop placed to the left of a lens is also its entrance pupil. Similarly, a stop placed to the right of a lens is also its exit pupil. Finally, a stop placed at a single thin lens is both its entrance and exit pupils.

1.3   Wave and Ray Aberrations

In this section, we define the wave aberration associated with a ray and relate it to its transverse ray aberration in an image plane. The *optical path length* of a ray in a medium of refractive index n is equal to n times its geometrical path length. If rays from a point object are traced through the system and up to the exit pupil such that each one travels an optical path length equal to that of the chief ray, the surface passing through their end points is called the system *wavefront* for the point object under consideration. If the wavefront is spherical with its center of curvature at the Gaussian image point, we say that the Gaussian image is perfect. If, however, the wavefront deviates from this *Gaussian spherical wavefront*, we say that the Gaussian image is aberrated. The optical deviations (i.e., geometrical deviations times the refractive index n of the image space) of the wavefront from a Gaussian spherical wavefront are called wave aberrations. The optical deviation of the wavefront along a certain ray from the Gaussian spherical wavefront is called the *wave aberration* of that ray. It represents the difference between the optical path lengths of the ray under consideration and the chief ray in traveling from the point object to the reference sphere. Accordingly, the wave aberration associated with the chief ray is zero. The wave aberration associated with a ray is positive if it has to travel an extra optical path length, compared to the chief ray, in order to reach the Gaussian spherical wavefront. The Gaussian spherical wavefront is also called the *Gaussian reference sphere*.

Figures 1-2 illustrate the reference sphere S and the aberrated wavefront W for on- and off-axis point objects whose Gaussian images lie at P_0' and P', respectively. The coordinate system is also illustrated in this figure. We choose a right-hand coordinate system such that the optical axis lies along the z axis. The object, entrance pupil, exit pupil, and the Gaussian image lie in mutually parallel planes that are perpendicular to this axis, with their origins lying along the axis. We assume that a point object

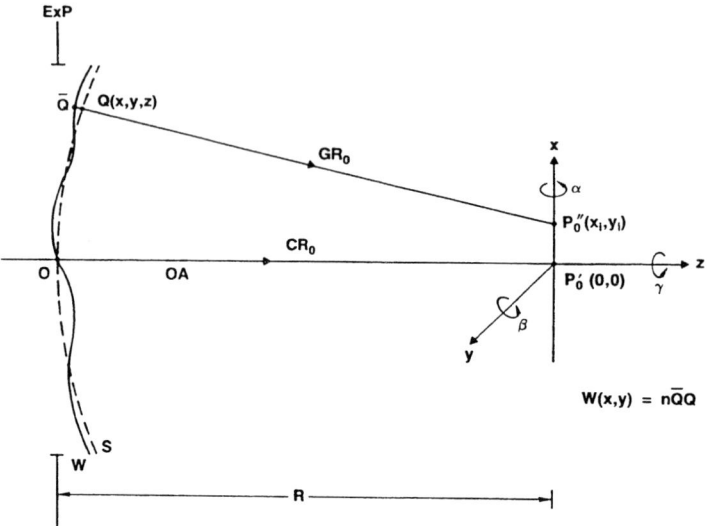

Figure 1–2a. Aberrated wavefront for an on-axis point object P_0 with a Gaussian image at P_0'. The reference sphere S of radius of curvature R is centered at the Gaussian image point. The wavefront W and reference sphere S pass through the center O of the exit pupil *ExP*. A general ray GR_0 is shown intersecting the Gaussian image plane at a point (x_i, y_i). The angular directions (α, β, γ) are also shown in the figure.

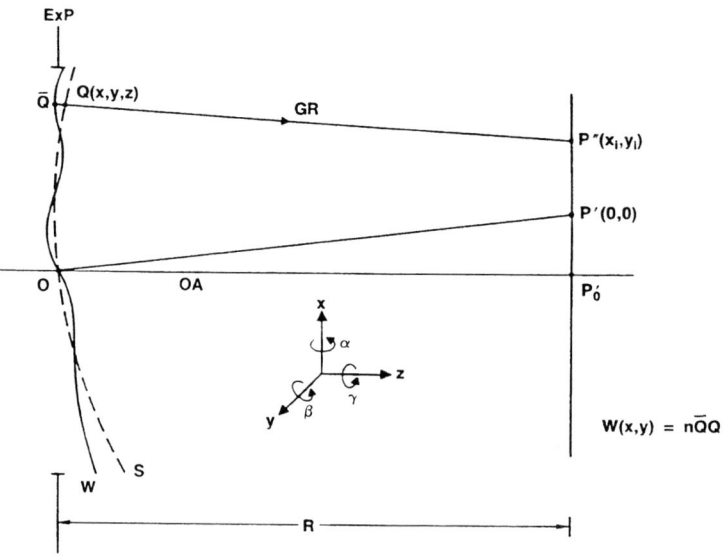

Figure 1–2b. Aberrated wavefront for an off-axis point object P with a Gaussian image at P'. The reference sphere S of radius of curvature R is centered at the Gaussian image point. The wavefront W and reference sphere S pass through the center O of the exit pupil *ExP*. A general ray GR is shown intersecting the Gaussian image plane at a point (x_i, y_i) with respect to the Gaussian image point.

such as P lies along the x axis. The zx plane containing the point object and the optical axis is called the *tangential* or the *meridional plane*. The Gaussian image P' lying in the Gaussian image plane along its x axis also lies in the tangential plane. This may be seen by a consideration of a tangential object ray and Snell's law according to which the incident and refracted or reflected rays at a surface lie in the same plane. The chief ray always lies in the tangential plane. The plane normal to the tangential plane but containing the chief ray is called the *sagittal plane*. As the chief ray bends when it is refracted or reflected by an optical surface, so does the sagittal plane.

Consider an image ray such as GR in Figure 1-2b passing through a point Q with coordinates (x,y,z) on the reference sphere of radius of curvature R centered at the Gaussian image point. We let $W(x,y)$ represent its wave aberration $n\overline{QQ}$, since z is related to x and y by virtue of Q being on the reference sphere. It can be shown that the ray intersects the Gaussian image plane at a point P'' whose coordinates with respect to the Gaussian image point P' are approximately given by

$$(x_i, y_i) = \frac{R}{n}\left(\frac{\partial W}{\partial x}, \frac{\partial W}{\partial y}\right) . \qquad (1\text{-}1)$$

[Equation (1-1) is derived in Born and Wolf and in Welford. Note however, that Welford uses a sign convention for the wave aberration that is opposite to ours.] The displacement $P'P''$ of a ray from the Gaussian image point is called its *geometrical* or *transverse ray aberration* and its coordinates (x_i, y_i) in the Gaussian image plane are called its ray aberration components. Since the rays are normal to a wavefront, the ray aberrations depend on the shape of the wavefront and, therefore, on its geometrical path difference from the reference sphere. The division of W by n in Eq. (1-1) converts the optical path length difference into geometrical path length difference. When an image is formed in free space, as is often the case in practice, then $n = 1$.

The distribution of rays from a point object in an image plane is called the *ray spot diagram*. (Such diagrams are discussed in Chapter 7.) When the wavefront is spherical with its center of curvature at the Gaussian image point, then the wave and ray aberrations are zero. In that case, all of the object rays transmitted by the system pass through the Gaussian image point, and the image is *perfect*. We shall refer to $W(x,y)$ as the wave aberration at a projected point (x,y) in the plane of the exit pupil. If (r,θ) represent the corresponding polar coordinates, they are related to the rectangular coordinates according to

$$(x,y) = r(\cos\theta, \sin\theta) . \qquad (1\text{-}2)$$

1.4 Defocus Aberration

We now discuss defocus wave aberration of a system and relate it to its longitudinal defocus. Consider an imaging system for which the Gaussian image of a point object is located at P_1. As indicated in Figure 1-3, let the wavefront for this point object be spherical with a center of curvature at P_2 (due to field curvature discussed in Section

1.6) such that P_2 lies on the line OP_1 joining the center O of the exit pupil and the Gaussian image point P_1. The aberration of the wavefront with respect to the Gaussian reference sphere is its optical deviation from it. This deviation is given by nQ_2Q_1, where n is the refractive index of the image space and Q_2Q_1, as indicated in the figure, is approximately equal to the difference in the sags of the reference sphere and the wavefront at a height r. (The *sag* of a surface at a certain point on it represents its deviation at that point along its axis of symmetry from a plane surface that is tangent to it at its vertex). Thus, the *defocus wave aberration* at a point Q_1 at a distance r from the optical axis is given by

$$W(r) = \frac{n}{2}\left(\frac{1}{R_1} - \frac{1}{R_2}\right)r^2 \quad , \tag{1-3a}$$

where R_1 and R_2 are the radii of curvature of spherical surfaces centered at P_1 and P_2, respectively, passing through the center of the exit pupil. We note that the defocus wave aberration is proportional to r^2. If $R_1 \simeq R_2 = R$, and $\Delta R = R_2 - R_1$, then Eq. (1-3a) may be written

$$W(r) = \frac{n}{2}(\Delta R/R^2)r^2 \quad . \tag{1-3b}$$

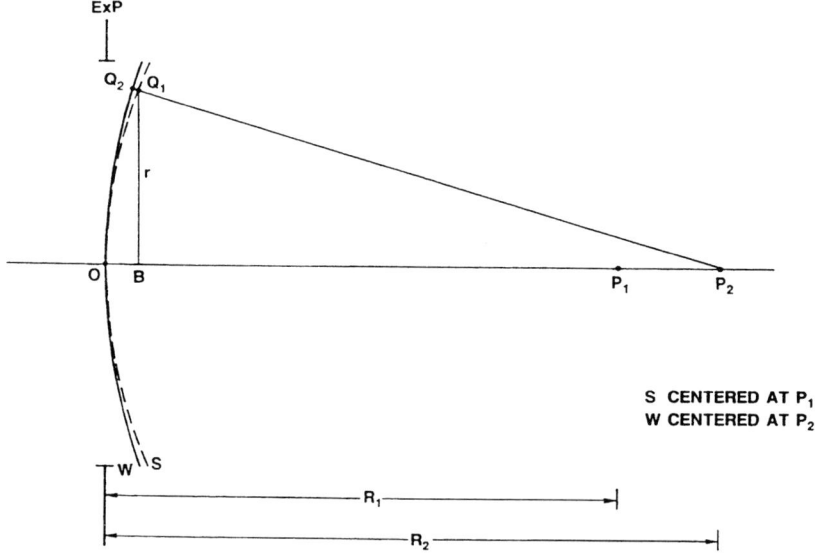

Figure 1–3. Defocused wavefront W is spherical with a radius of curvature R_2 centered at P_2. The reference sphere S with a radius of curvature R_1 is centered at P_1. Both W and S pass through the center of the exit pupil *ExP*. The ray Q_2P_2 is normal to the wavefront at Q_2. OB represents the sag of Q_1.

The quantity ΔR is called the *longitudinal defocus*. The ray aberrations corresponding to a defocus wave aberration are discussed in Chapter 7.

A defocus aberration is also introduced if the image is observed in a plane other than the Gaussian image plane. Consider, for example, an imaging system forming an aberration-free image at the Gaussian image point P_2 (note that the Gaussian image is now located at P_2). Thus, the wavefront at the exit pupil is spherical passing through its center O with its center of curvature at P_2. Let the image be observed in a defocused plane passing through a point P_1 which lies on the line joining O and P_2. For the observed image at P_1 to be aberration free, the wavefront at the exit pupil must be spherical with its center of curvature at P_1. Such a wavefront forms the reference sphere with respect to which the aberration of the actual wavefront must be defined. The aberration of the wavefront at a point Q_1 on the reference sphere is given by Eqs. (1-3).

If the exit pupil is circular with a radius a, then Eq. (1-3b) may be written

$$W(\rho) \;=\; (n\Delta R/8F^2)\rho^2 \tag{1-3c}$$

$$= \; A_d\rho^2, \tag{1-3d}$$

where $F = R/2a$ is the *focal ratio* or the *f-number* of the image-forming light cone, $\rho = r/a$ is the normalized distance of a point in the plane of the pupil from its center, and $A_d = n\Delta R/8F^2$ is the peak value of the defocus aberration. Note that a positive value of A_d implies a positive value of ΔR. Thus, an imaging system having a positive value of defocus aberration A_d can be made defocus free if the image is observed in a plane lying farther from the plane of the exit pupil, compared to the defocused image plane, by a distance $8A_dF^2/n$. Similarly, a positive defocus aberration of $A_d = n\Delta R/8F^2$ is introduced into the system if the image is observed in a plane lying closer to the plane of the exit pupil, compared to the defocus-free image plane, by a distance ΔR.

1.5   Wavefront Tilt

Now we describe the relationship between a wavefront tilt and the corresponding tilt aberration. As indicated in Figure 1-4, consider a spherical wavefront centered at P_2 in the Gaussian image plane passing through the Gaussian image point P_1. The wave aberration of the wavefront at Q_1 is its optical deviation nQ_2Q_1 from a reference sphere centered at P_1. It is evident that, for small values of the ray aberration P_1P_2, the wavefront and the reference sphere are tilted with respect to each other by an angle β. The wavefront tilt may be due to distortion discussed in Section 1.6 and/or due to an inadvertently tilted element of the imaging system. The ray and the wave aberrations can be written

$$x_i \;=\; R\beta \tag{1-4}$$

and

$$W(r,\theta) = n\beta r\cos\theta \quad , \tag{1-5a}$$

respectively, where $P_1 P_2 = x_i$ and (r,θ) are the polar coordinates of point Q_1.

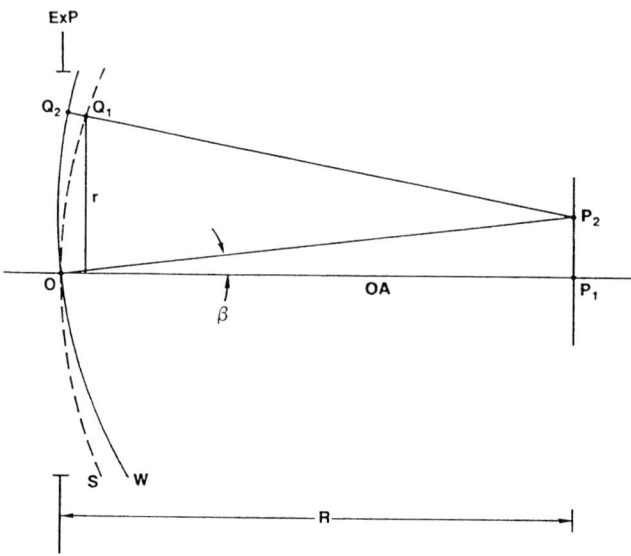

Figure 1–4. Wavefront tilt. The spherical wavefront W is centered at P_2 while the reference sphere S is centered at P_1. Thus, for small values of $P_1 P_2$, the two spherical surfaces are tilted with respect to each other by an angle $\beta = P_1 P_2/R$, where R is their radius of curvature. The ray $Q_2 P_2$ is normal to the wavefront at Q_2.

Once again, if the exit pupil is circular with a radius a, then letting $\rho = r/a$, Eq. (1-5a) may be written

$$W(\rho,\theta) = na\beta\rho\cos\theta \tag{1-5b}$$

$$= A_t\,\rho\cos\theta \quad , \tag{1-5c}$$

where $A_t = na\beta$ is the peak value of the tilt aberration. Note that a positive value of A_t implies that the wavefront tilt angle β is also positive. Thus, if an aberration-free wavefront is centered at P_2, then observation with respect to P_1 as the origin implies that we have introduced a tilt aberration of $A_t\rho\cos\theta$.

1.6 Aberration Function of a Rotationally Symmetric System

The aberration function $W(r,\theta;h')$ of an optical imaging system with an axis of rotational symmetry depends on the object height h or the image height h' from the optical axis, and pupil coordinates (r,θ) of a point in the plane of the exit pupil, through three *rotational invariants* h'^2, r^2, and $h'r\cos\theta$. The aberration terms of degree 4 in the rectangular coordinates of the object and pupil points are called *primary aberrations*. Thus, the *primary aberration function* consists of a sum of five terms, e.g.,

$$W(r, \theta; h') = {}_0a_{40}\, r^4 + {}_1a_{31}\, h'r^3\cos\theta + {}_2a_{22}\, h'^2 r^2\cos^2\theta + {}_2a_{20}\, h'^2 r^2 + {}_3a_{11}\, h'^3 r\cos\theta \ , \quad (1\text{-}6)$$

where the subscripts of the aberration coefficients ${}_i a_{jk}$ represent the powers of h', r, and $\cos\theta$, respectively. Note that there is no term in h'^4 since the aberration of the chief ray ($r = 0$) must be zero. Since the wave aberration W has dimensions of length, the dimensions of the coefficients ${}_i a_{jk}$ are inverse length cube. The *order* of an aberration term is equal to the sum of the powers of h' and r; i.e., it is equal to its degree in the (x,y) coordinates of the object (or its image) and pupil points. It is equal to 4 for a primary aberration. Accordingly, the primary aberrations are called *fourth-order wave aberrations*. They are also called the *Seidel aberrations*. Since the ray aberrations are related to the wave aberrations by a spatial derivative [see Eq. (1–1)], their degree is lower by one. Accordingly, primary aberrations are also called *third-order ray aberrations*. The coefficients ${}_0a_{40}$, ${}_1a_{31}$, ${}_2a_{22}$, ${}_2a_{20}$, and ${}_3a_{11}$ represent the coefficients of *spherical aberration, coma, astigmatism, field curvature*, and *distortion*, respectively.

From Eq. (1-6) we note that only spherical aberration is independent of the object or image height. The field curvature, in its dependence on the pupil coordinates (r,θ), is like the defocus aberration discussed in Section 1.4. However, field curvature represents a defocus aberration that depends on the field h', thus requiring a curved image surface for its elimination. On the other hand, pure defocus aberration, such as that produced by observing the image in a plane other than the Gaussian image plane, is independent of the field h'. Similarly, distortion depends on the pupil coordinates as a wavefront tilt. However, distortion depends on the field as h'^3, but wavefront tilt produced by a tilted element in the system would be independent of h'.

For simplicity, we will use the notation a_s, a_c, a_a, a_d (d for defocus), and a_t (t for tilt) to represent the coefficients of spherical aberration, coma, astigmatism, field curvature, and distortion, respectively. For an optical system with a circular pupil of radius a, we can use the normalized radial variable $\rho = r/a$, suppress the explicit dependence on image height h', and write the primary aberration function in the form

$$W(\rho, \theta) = A_s\rho^4 + A_c\rho^3\cos\theta + A_a\rho^2\cos^2\theta + A_d\rho^2 + A_t\rho\cos\theta \ , \quad (1\text{-}7)$$

where A_i are the *peak aberration coefficients* given by

$$A_s = a_s a^4, \ A_c = a_c h'a^3, \ A_a = a_a h'^2 a^2, \ A_d = a_d h'^2 a^2, \ A_t = a_t h'^3 a \ . \quad (1\text{-}8)$$

It should be clear that, since $0 \le \rho \le 1$ and $0 \le \theta < 2\pi$, a peak aberration coefficient A_i, as the name implies, represents the maximum value of the corresponding aberration. This value occurs at a point on the edge of the pupil, i.e., for a marginal ray. With the image height suppressed, the defocus and tilt coefficients A_d and A_t are similar to the corresponding coefficients considered in Sections 1.4 and 1.5.

1.7   Effect of Change in Aperture Stop Position on the Aberration Function

Now we consider how the primary aberration function of a system changes due to a change in the position of its aperture stop. We remind ourselves that the wave aberration associated with a ray represents the difference between its optical path length and that of the chief ray in traveling from a point object to the reference sphere. Moreover, the chief ray is that object ray which passes through the center of the aperture stop. Hence, since the chief ray changes as the position of the aperture stop is changed, the wave aberration of a ray also changes.

Consider, as indicated in Figure 1-5, an optical imaging system forming an image P' of an off-axis point object, at a height h' from the optical axis. Let the aperture stop of the system be located at a position such that its exit pupil is located at ExP_1. Let the primary aberration function of the system be given by $W_{Q1}(x_1, y_1; h')$ representing the aberration of an image forming ray passing through a point (x_1, y_1) in the plane of the exit pupil with respect to the chief ray O_1P' passing through the center O_1 of the exit pupil.

Now, suppose we move the aperture stop to a new position along the optical axis such that the corresponding new exit pupil is located at ExP_2 with its center at O_2. A change in the stop position does not change the position of the image P'. Let L_1 and L_2 be the axial distances of the Gaussian image plane from the planes of the exit pupils ExP_1 and ExP_2, respectively. The chief ray $O_1 P'$ (or its extension) intersects the plane of exit pupil ExP_2 at O_2' with rectangular coordinates $(x_0, 0)$, where from similar triangles $O_1 O_2 O_2'$ and $O_1 P_0' P'$ one finds that

$$x_0 = \frac{h'}{L_1}(L_1 - L_2) \quad . \tag{1-9}$$

Note that the y coordinate is zero because it lies on the chief ray, which in turn lies in the tangential plane zx.

The aberration of a ray $Q_1 P'$ with respect to the chief ray $O_1 P'$ represents the aberration at a point Q_1 with respect to the aberration at O_1 (which is zero by definition). It is also equal to the aberration of the ray $Q_1 P'$ at Q_2 with respect to the aberration at O_2', where Q_2 represents the point of intersection of the ray $Q_1 P'$ with the plane of the exit pupil ExP_2. It is evident from the geometry of Figure 1-5 that

$$(x_1, y_1) = \frac{L_1}{L_2}[(x_2 - x_0), y_2] \quad , \tag{1-10}$$

where (x_2, y_2) are the coordinates of Q_2 with respect to O_2 as the origin. Thus, the aberration at Q_2 with respect to its value at O_2' may be obtained by substituting Eq. (1-10) into the expression for $W_{Q1}(x_1, y_1)$, i.e.,

$$W_{Q2}(x_2, y_2) = W_{Q1}\left(\frac{L_1}{L_2}[(x_2 - x_0), y_2]\right) \quad . \tag{1-11}$$

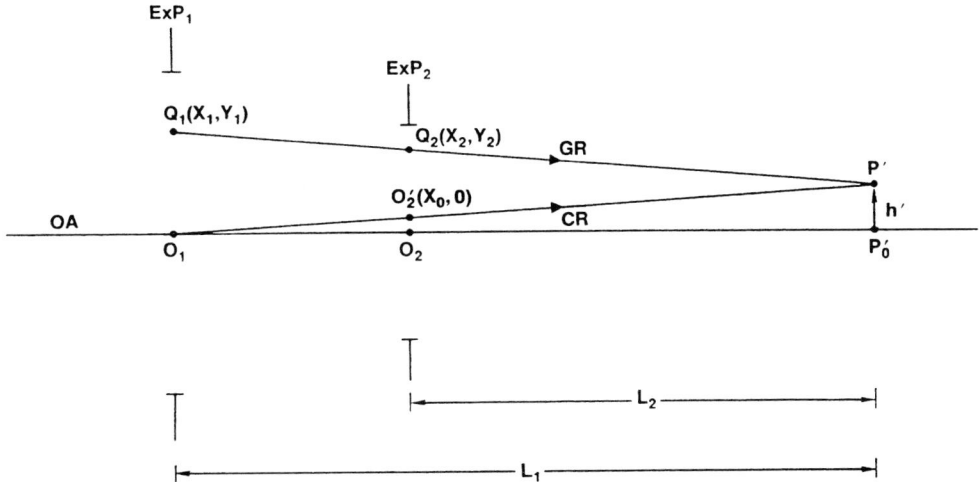

Figure 1–5. Exit pupils *ExP*₁ and *ExP*₂ corresponding to two positions of the aperture stop of an optical system forming Gaussian image *P'* of an off-axis point object. The chief ray *CR* shown is for the exit pupil *ExP*₁.

Note that the aberration function referred to the new exit pupil is zero at $(x_0, 0)$. In order that the aberration at the center O_2 of the new exit pupil be zero, we define a new aberration function $W(x_2, y_2; h')$ with respect to the new chief ray O_2P' (not shown in Figure 1-5), i.e.,

$$W(x_2, y_2; h') = W_{Q2}(x_2, y_2) - W_{Q2}(0,0)$$

$$= W_{Q1}\left(\frac{L_1}{L_2}[(x_2 - x_0), y_2]\right) - W_{Q1}(-x_0 L_1/L_2, 0) \quad . \qquad (1\text{-}12)$$

Let the primary aberration function at ExP_1 be given by

$$W_{Q1}(x_1, y_1; h') = a_{s1}(x_1^2 + y_1^2)^2 + a_{c1}h'x_1(x_1^2 + y_1^2)$$

$$+ a_{a1}h'^2 x_1^2 + a_{d1}h'^2(x_1^2 + y_1^2) + a_{t1}h'^3 x_1 \quad . \qquad (1\text{-}13)$$

Substituting Eq. (1-13) into Eq. (1-12) and noting from Figure 1-5 that the ratio of the radii of the two exit pupils is equal to the ratio of their distances from the Gaussian image plane, we can show that the old and the new peak aberration coefficients are related to each other according to

$$A_{s2} = A_{s1} \tag{1-14a}$$

$$A_{c2} = A_{c1} - 4bA_{s1} \tag{1-14b}$$

$$A_{a2} = A_{a1} - 2bA_{c1} + 4b^2A_{s1} \tag{1-14c}$$

$$A_{d2} = A_{d1} - bA_{c1} + 2b^2A_{s1} \tag{1-14d}$$

and

$$A_{t2} = A_{t1} - 2b(A_{a1} + A_{d1}) + 3b^2A_{s1} - 4b^3A_{s1} \quad , \tag{1-14e}$$

where

$$b = (L_1 - L_2)h'/a_1L_2 \quad . \tag{1-15}$$

In Eq. (1-15), a_1 is the radius of the exit pupil ExP_1. It is evident from Eqs. (1-14) that, because of a shift in the position of the aperture stop, an aberration of a certain order in pupil coordinates introduces aberrations of all lower orders as well. For example, a term in spherical aberration not only gives spherical aberration, but introduces coma, astigmatism, field curvature, and distortion as well.

From Eq. (1-14a), we note that the peak spherical aberration of a system is independent of the position of its aperture stop. Equation (1-14b) shows that if a system is free of spherical aberration, then the peak value of its coma is independent of the position of its aperture stop. It also shows that if spherical aberration is not zero, its coma can be made zero by selecting an aperture stop position corresponding to

$$b = \frac{A_{c1}}{4A_{s1}} \quad or \quad \frac{L_1}{L_2} = 1 + \frac{a_{c1}}{4a_{s1}} \quad . \tag{1-16}$$

Similarly, Eqs. (1-14c) and (1-14d) show that if a system is free of spherical aberration and coma, then the peak values of its astigmatism and field curvature are independent of the position of its aperture stop. Finally, Eq. (1-14e) shows that the peak value of distortion depends on the position of the aperture stop unless spherical aberration, coma, and the sum of astigmatism and field curvature are each zero.

It should be noted that the optical path length of a ray, or its optical path length difference with respect to another, does not change with a change in the position of the aperture stop. However, since the chief ray does change, the new aberration function merely describes the wave aberrations of rays with respect to the new chief ray. The position of the aperture stop also affects which and how many of the object rays are transmitted by the system. Indeed, for high-quality imaging systems, a lens designer chooses the position of the aperture stop judiciously so that rays with large aberrations are blocked by it without a substantial loss in the amount of transmitted light.

An example of the utility of an appropriate position of the aperture stop is considered in Chapter 4, where it is shown that a spherical mirror with aperture stop located at its center of curvature suffers only from spherical aberration and field curvature. Coma, astigmatism, and distortion, which may be present for any other position of the aperture stop, are identically zero for this specific location. Indeed, such a location of the aperture stop forms the basis of the *Schmidt camera* discussed in Chapter 5.

1.8   Aberrations of a Spherical Refracting Surface

In this section, we discuss imaging by a spherical refracting surface. We give equations for Gaussian imaging and expressions for its primary aberrations for an arbitrary position of its aperture stop. The results given here form the cornerstone for imaging by more complicated systems. By making simple but appropriate changes in them, the results for a spherical mirror can be obtained immediately, as indicated in Chapter 4.

As illustrated in Figure 1-6, consider a spherical refracting surface of radius of curvature R separating media of refractive indices n and n'. The line joining its vertex V_0 and its center of curvature C is called the optical axis. Consider a point object P at a distance S from the vertex and at a height h from the optical axis. Let P' be its Gaussian image at a distance S' and at a height h'.

The relationships between the distances and heights of the object and image points are given by Gaussian optics according to

$$\frac{n}{S} + \frac{n'}{S'} = \frac{n' - n}{R} \tag{1-17a}$$

$$= \frac{n}{f} = \frac{n'}{f'} \tag{1-17b}$$

and

$$M = \frac{h'}{h} = -\frac{S' - R}{S + R} \tag{1-18a}$$

$$= -\frac{nS'}{n'S} \quad , \tag{1-18b}$$

where f and f' are the left and the right *focal lengths* of the refracting surface and M is the *transverse magnification* of the image. Here f represents the object distance S such that the image distance S' is infinity. Similarly, f' represents the image distance S' such that the object distance S is infinity. The negative sign in Eqs. (1-18) indicates that the height of an object or image below the optical axis is considered numerically negative. (See the Appendix for sign convention.)

Let the aperture stop be located at a distance L from the image as indicated in Figure 1-6. It is evident from the geometry of the figure that the aperture stop is also the exit pupil of the imaging system. For a refracting surface, L is positive when the image lies to the right of the exit pupil.

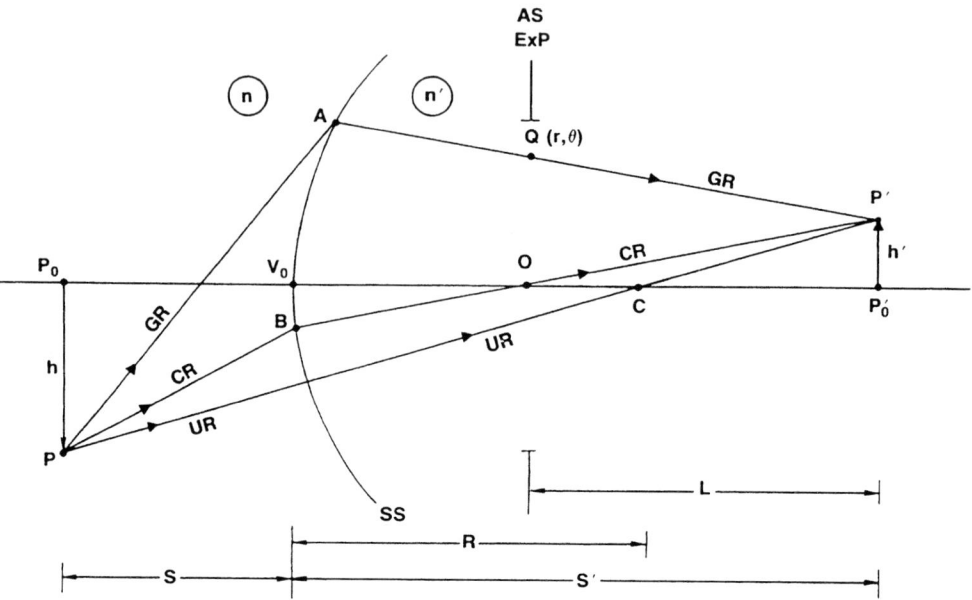

Figure 1–6. Imaging by a spherical refracting surface *SS* of radius of curvature *R* having its center of curvature at *C*, separating media of refractive indices *n* and *n'*. The aperture stop and, therefore, the exit pupil are located at a distance *L* from the Gaussian image. The undeviated ray *UR* helps locate the Gaussian image *P'* of a point object *P*.

The ray *PBP'* passing through the center *O* of the aperture stop, which is also the exit pupil, is called the chief ray for the point object *P*. The aberration of a ray *PAP'* incident at a point *A* on the refracting surface and passing through a point *Q* in the plane of the exit pupil with polar coordinates (r, θ) with respect to the chief ray *PBP'* is given by

$$W(A) = [PAP'] - [PBP'] \quad ,$$

where the square brackets indicate an optical path length. It should be noted that the rays *PA* and *PB* from the point object *P* incident at points *A* and *B*, respectively, on the refracting surface, may not pass through the Gaussian image point *P'* after refraction unless the image at *P* is aberration free. It can be shown that, up to the fourth order in pupil and object or image coordinates, the aberration $W(A) \equiv W(Q)$ reduces to

$$W_s(r, \theta; h') = a_{ss}r^4 + a_{cs}h'r^3\cos\theta + a_{as}h'^2r^2\cos^2\theta + a_{ds}h'^2r^2 + a_{ts}h'^3r\cos\theta \quad , \quad (1\text{-}19)$$

where

$$a_s = -\frac{n'(n'-n)}{8n^2}\left(\frac{1}{R}-\frac{1}{S'}\right)^2\left(\frac{n'}{R}-\frac{n+n'}{S'}\right) \tag{1-20}$$

$$a_{ss} = (S'/L)^4 a_s \tag{1-21a}$$

$$a_{cs} = 4d a_{ss} \tag{1-21b}$$

$$a_{ds} = 4d^2 a_{ss} \tag{1-21c}$$

$$a_{ds} = 2d^2 a_{ss} - \frac{n'(n'-n)}{4nRL^2} \tag{1-21d}$$

$$a_{ts} = 4d^3 a_{ss} - \frac{n'(n'-n)d}{2nRL^2} \tag{1-21e}$$

and

$$d = \frac{R-S'+L}{S'-R} \quad. \tag{1-22}$$

Note that L is (approximately) the radius of curvature of the reference sphere passing through the center of the exit pupil with its center of curvature at P'. Equation (1-19) gives the wave aberration at a point (r,θ) in the plane of the exit pupil for a point object whose Gaussian image height is h'.

The second term on the right-hand side of Eq. (1-21d) may be called the *coefficient of Petzval curvature*, and we denote it by a_p, i.e.,

$$a_p = -\frac{n'(n'-n)}{4nRL^2} \quad. \tag{1-23}$$

The corresponding wave aberration may be written

$$W_p(r) = a_p h'^2 r^2 \quad. \tag{1-24}$$

This aberration reduces to zero if the image is observed at a (longitudinal) distance ΔL from the Gaussian image, where ΔL is related to the aberration according to Eq. (1-3b), i.e.,

$$W_p(r) = \frac{n'}{2}\frac{\Delta L}{L^2}r^2 \quad. \tag{1-25}$$

If the image is observed on a spherical surface of radius of curvature R_p passing through the axial image point P_0', the longitudinal defocus ΔL for a Gaussian image at a height h' is given by its sag

$$\Delta L = h'^2/2R_p \quad. \tag{1-26}$$

Substituting Eq. (1-26) into Eq. (1-25), and Eq. (1-23) into Eq. (1-24), and equating the results obtained, we find that

$$R_p = \frac{nR}{n - n'} \ . \tag{1-27}$$

We note that R_p, called the *Petzval radius of curvature*, is independent of the object position. The image surface under consideration is called the *Petzval image surface*. From Eqs. (1-21c), (1-21d), and (1-27), we may write

$$2a_{ds} - a_{as} = n'/2R_pL^2 \ . \tag{1-28}$$

We will utilize Eq. (1-28) in Section 7.7 where we relate the Petzval surface to the sagittal and tangential image surfaces which result from astigmatism.

Letting $h' = 0$ in Eq. (1-19), we note that the image P_0' of an axial point object P_0 suffers from spherical aberration only. The amount of spherical aberration does not change as we move from an on-axis to an off-axis point object. Note that when the aperture stop and, therefore, the exit pupil are located at the refracting surface, then $L = S'$ and Eqs. (1-21a) and (1-22) reduce to $a_{ss} = a_s$ and $d = R/(S' - R)$, respectively.

It is evident from Eq. (1-20) that $a_s = 0$ when $S' = (n + n')R/n$, which in turn corresponds to $S = -(n + n')R/n'$. Accordingly, a_{ss}, a_{cs} and a_{as} are all zero. Two conjugate points for which spherical aberration, coma, and astigmatism are zero are called *anastigmatic*. Depending on whether R is positive or negative, the object or the image point is virtual for these anastigmatic points. We note that spherical aberration is also zero when $S' = R$ and $S = -R$. However, in this case, coma is also zero, but astigmatism is not due to the d^2 factor on the right–hand side of Eq. (1-21c). Two conjugate points for which spherical aberration and coma are zero are called *aplanatic*. Thus, the points under consideration are aplanatic, and, once again, either the object or the image is virtual.

1.9 Aberration Function of a Multielement System

Consider an optical system made up of a series of coaxial refracting and/or reflecting surfaces. Each surface produces primary aberrations with its own value of h' and L. The image of a point object formed by the first surface acts as an object for the second surface, and so on. The aberration is calculated surface by surface, and the aberration of the system is obtained by adding the aberration contributions of all the surfaces. Since the aberration of a surface is calculated at a point on its exit pupil, the coordinates of a pupil point must be transformed using *pupil magnification* of a surface to obtain the aberration contribution of a surface at a point on the exit pupil of the system. Similarly, image magnification of a surface can be used to obtain the system aberration in terms of the height of the image formed by the system.

For example, if $W_1(x_1,y_1;h_1')$ represents the aberration at a point (x_1,y_1) in the plane of the exit pupil of the first surface for an image of height h_1', it can be converted to an aberration contribution at a point (x_2,y_2) in the plane of the exit pupil of the second surface and image height h_2' by letting $(x_1, y_1; h_1') = (x_2/m_2, y_2/m_2; h_2'/M_2)$, where m_2 and M_2 represent the pupil and image magnifications, respectively, for the second surface. Thus, if $W_2(x_2,y_2;h_2')$ represents the aberration contribution of the second surface at a point (x_2,y_2) in the plane of its exit pupil corresponding to an image height of h_2', the total aberration for the two surfaces will be given by

$$W_s(x_2,y_2;h_2') = W_1\left(\frac{x_2}{m_2}, \frac{y_2}{m_2}; \frac{h_2'}{M_2}\right) + W_2(x_2,y_2;h_2') \quad . \tag{1-29}$$

This process can be continued to obtain the system aberration $W(x,y;h')$ at a point (x,y) in the plane of the exit pupil of the system corresponding to a height h' of the image of a point object formed by the system. It is utilized, for example, to calculate the aberrations of a thin lens in Chapter 2 and a plane-parallel plate in Chapter 3.

Since the refractive index of a transparent substance varies with optical wavelength, the angle of refraction of a ray also varies with it. Hence, even the Gaussian image of a multiwavelength point object formed by a refracting system is generally not a point. The distance and height of the image vary with the wavelength. The axial and transverse extents of the image are called *longitudinal* and *transverse chromatic aberrations*, respectively. They describe the chromatic change in position and magnification of the image, respectively. The monochromatic aberrations of a refracting system also vary with the wavelength, but such a variation is small for a small change in the wavelength and is usually negligible.

Appendix: Sign Convention

1. Object is placed to the left of an optical system so that, initially, light travels from left to right.

2. Radius of curvature R of a surface is positive if its center of curvature lies to the right of its vertex.

3. Distance S of an object from (the vertex of) a surface is positive if the object lies to the left of the surface.

4. Distance S' of an image from (the vertex of) a refracting surface is positive if the image lies to the right of the surface. Distance of an image from (the vertex of) a reflecting (mirror) surface is positive if the image lies to the left of the surface.

5. Height h of an object or h' of an image from the optical axis is positive if it lies above the axis.

6.   Wave aberration $W(Q)$ of a wavefront at a point Q on a reference sphere, which represents the optical deviation between the two along an object ray passing through Q, is positive if the ray has to travel an extra optical path length to reach the reference sphere. The wavefront and the reference sphere pass through the center of the exit pupil so that the aberration of the chief ray is zero.

7.   Distance L between an image and a corresponding exit pupil is positive if the image lies to the right of the exit pupil in the case of a refracting surface. In the case of a reflecting surface, L is positive if the image lies to the left of the exit pupil.

In Figure 1-6, quantities R, S, S', h' are all numerically positive. Only, h is negative.

CHAPTER 2

Aberrations of a Thin Lens

2.1 Introduction

Among the simple optical imaging systems, a thin lens consisting of two spherical surfaces is the most common as well as practical. By applying the results of Section 1.8 and the procedure of Section 1.9, we give the imaging equations and expressions for the primary aberrations of a thin lens with aperture stop located at the lens. Its aberrations for other locations of the aperture stop may be obtained by applying the results of Section 1.7 to those given here. It is shown that when both an object and its image are real, the spherical aberration of a thin lens cannot be zero (unless its surfaces are made nonspherical). We illustrate by a numerical example, however, that it is possible to design a two-lens combination such that its spherical aberration and coma are both zero. In such a combination, these aberrations associated with one lens cancel the corresponding aberrations of the other. This cancellation is illustrated with a numerical example.

2.2 Gaussian Imaging

Consider a thin lens of refractive index n and focal length f consisting of two spherical surfaces of radii of curvature R_1 and R_2 as illustrated in Figure 2-1. A lens is considered *thin* if its thickness is negligible compared to f, R_1, and R_2. Its optical axis OA is the line joining the centers of curvature C_1 and C_2 of its surfaces. Since the lens is thin, we neglect the spacing between its surfaces. We assume that its aperture stop AS is located at the lens, so that its entrance and exit pupils EnP and ExP, respectively, are also located there. The lens is located in air; therefore, the refractive index of the surrounding medium is 1.

Consider a point object P located at a distance S from the lens and at a height h from its axis. The first surface forms the image of P at P' and the second surface forms the image of P' at P''. Applying the results of Section 1.8 to imaging by the two surfaces of the lens, where $n = 1$ and $n' = n$ for the first surface and $n = n$ and $n' = 1$ for the second surface, we can show that the image distance S' and its height h' are given by the relations

$$\frac{1}{S} + \frac{1}{S'} = (n-1)\left(\frac{1}{R_1} - \frac{1}{R_2}\right) \tag{2-1a}$$

$$= \frac{1}{f} \tag{2-1b}$$

and

$$M = \frac{h'}{h} \tag{2-2a}$$

$$= -\frac{S'}{S} , \tag{2-2b}$$

21

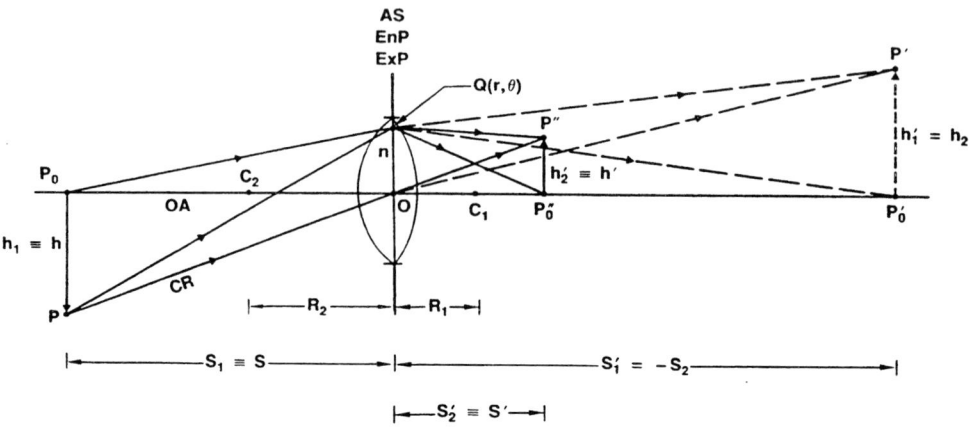

Figure 2–1. Imaging by a thin lens of refractive index n formed by two surfaces of radii of curvature R_1 and R_2 with centers of curvature at C_1 and C_2. Whereas R_1 is numerically positive, R_2 is negative. $P_0'P'$ is the Gaussian image of object P_0P formed by the first surface. $P_0''P''$ is the image of virtual object $P_0'P'$ formed by the second surface. The height h of the point object P is numerically negative since it lies below the optical axis OA. The height h' of the corresponding final image P'' is numerically positive, since it lies above the optical axis. For a thin lens, its thickness is neglected. The aperture stop AS, entrance pupil EnP, and the exit pupil ExP are all located at the lens.

respectively, where M is the magnification of the image. Note that we are able to write Eq. (2-1b) because by definition, the *focal length f* is the image distance when the object is at infinity.

2.3   Primary Aberrations

The aberration of an object ray PQP'' passing through a point Q in the plane of the exit pupil with polar coordinates (r,θ) with respect to the chief ray POP'' passing through the center O of the exit pupil is given by

$$W(Q) = [PQP''] - [POP''] .$$

Noting that the optical path lengths $[P'Q]$ and $[P'O]$ are negative since they are *virtual*, the aberration of the ray can be written in terms of the aberrations produced by the two surfaces, i.e.,

$$W(Q) = \{[PQP'] - [POP']\} + \{[P'QP''] - [P'OP'']\} \ .$$

By applying the results of Section 1.8 to the two surfaces of the thin lens and following the procedure of Section 1.9, it can be shown that the primary aberration function of the thin lens is given by

$$W_l(r, \theta; h') = a_s r^4 + a_c h' r^3 \cos\theta + a_a h'^2 r^2 \cos^2\theta + a_d h'^2 r^2 \ , \tag{2-3}$$

where

$$a_s = -\frac{1}{32n(n-1)f^3}\left[\frac{n^3}{n-1} + (3n+2)(n-1)p^2 + \frac{n+2}{n-1}q^2 + 4(n+1)pq\right] \tag{2-4a}$$

$$a_c = \frac{1}{4nf^2 S'}\left[(2n+1)p + \frac{n+1}{n-1}q\right] \tag{2-4b}$$

$$a_a = -1/2fS'^2 \tag{2-4c}$$

and

$$a_d = -(n+1)/4nfS'^2 \ . \tag{2-4d}$$

Note that there is no distortion term in Eq. 2-3; i.e., a thin lens with an aperture stop at the lens does not produce any distortion. The quantities p and q are called the *position* and *shape factors* of a thin lens, respectively. They are given by

$$p = (2f/S) - 1 \tag{2-5a}$$

$$= 1 - 2f/S' \tag{2-5b}$$

and

$$q = (R_2 + R_1)/(R_2 - R_1) \ . \tag{2-6}$$

Several examples of the position and shape factors are illustrated in Figures 2-2 and 2-3, respectively. Both positive and negative lenses (in the sense of the sign of their focal length) are considered in these figures. The names associated with the different lens shapes are also noted in Figure 2-3.

We note from Eqs. (2-4c) and (2-4d) that astigmatism and field curvature coefficients of a thin lens do not depend on its position and shape factors. Moreover, the astigmatism coefficient does not depend on its refractive index, and the field curvature coefficient is smaller than the astigmatism coefficient by a factor of $(n+1)/2n$.

2.4   Spherical Aberration and Coma

From Eqs. (2-4a) and (2-4b) we note that the spherical aberration and coma of a thin lens depend on its position and shape factors. For a given position factor, the value of the shape factor which minimizes the spherical aberration is given by the condition

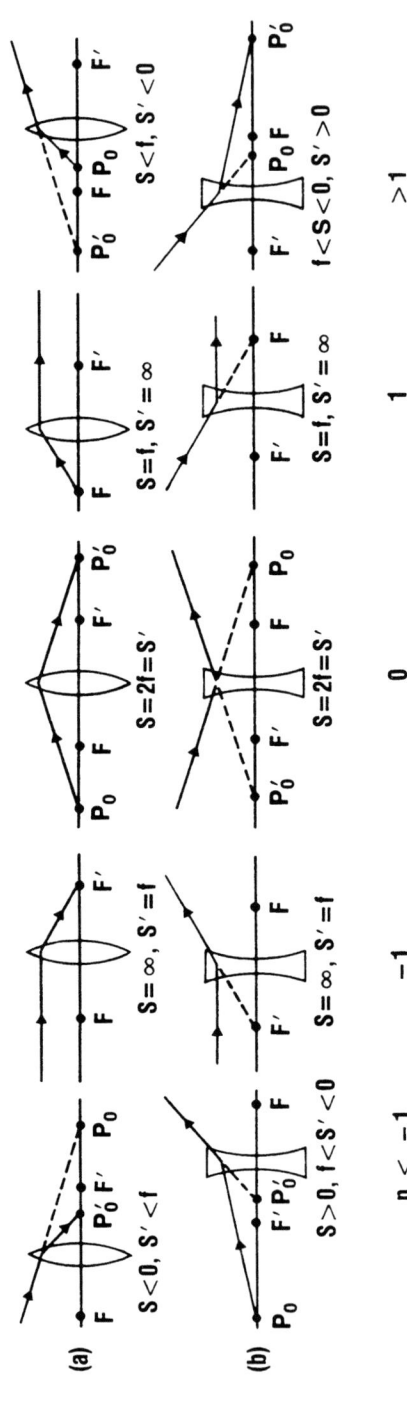

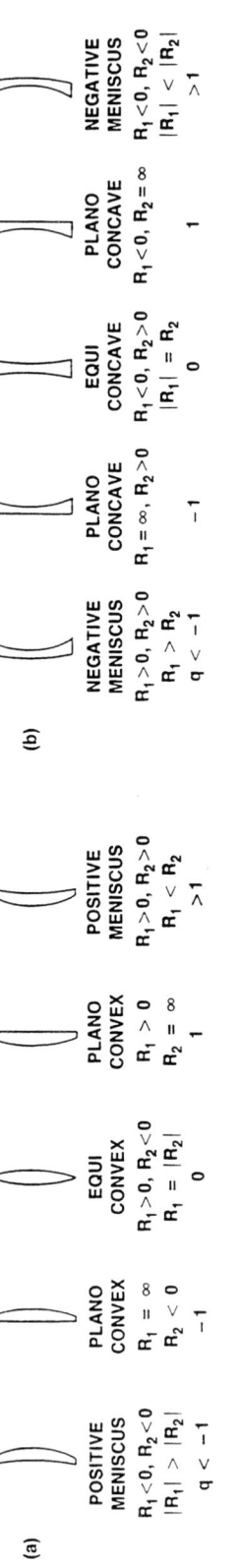

Figure 2–2. Position factor $1 < p < -1$ of a thin lens. (a) Positive lens. (b) Negative lens. F and F' are the focal points of a lens of focal length f. P_0 and P_0' represent an axial point object and its corresponding image, respectively. S and S' are the object and image distances, respectively, from the center of the lens. When the object (image) is at infinity, the image (object) is at the corresponding focal point.

Figure 2–3. Shape factor $1 < q < -1$ of a thin lens with surfaces of radii of curvature R_1 and R_2. (a) Positive lens. (b) Negative lens.

$$\frac{\partial a_s}{\partial q} = 0 \; . \tag{2-7}$$

Thus, we obtain

$$q_{min} = -2p\frac{n^2-1}{n+2} \; . \tag{2-8}$$

Substituting Eq. (2-8) into Eq. (2-4a), we obtain the corresponding *minimum spherical aberration*

$$a_{smin} = -\frac{1}{32f^3}\left[\left(\frac{n}{n-1}\right)^2 - \frac{n}{n+2}p^2\right] \; . \tag{2-9}$$

Thus, following Eq. (2-4a), we note that, for a given value of p, a_s as a function of q follows a parabola with a vertex lying at (q_{min}, a_{smin}). For different values of p, the parabolas have the same shape but different vertices. It is evident from Eqs. (2-5) that when both an object and its image are real

$$-1 \le p \le 1, \; or \; p^2 \le 1 \; . \tag{2-10}$$

As indicated in Figure 2-2, the case $p = -1$ corresponds to an object at infinity and the image at the focal plane of the lens. Similarly, $p = 1$ corresponds to an object at the focal plane and the image at infinity. The case $p = 0$ corresponds to object and image lying at a distance $2f$ on each side of the lens, respectively. For spherical aberration to be zero, Eq. (2-9) yields

$$p^2 = \frac{n(n+2)}{(n-1)^2}$$

$$> 1 \; . \tag{2-11}$$

Hence, spherical aberration of a thin lens cannot be zero when both the object and its image are real.

For a thin lens with a refractive index $n = 1.5$, Eqs. (2-4a), (2-8), and (2-9) reduce to

$$a_s = -\frac{1}{24f^3}(6.75 + 3.25p^2 + 7q^2 + 10pq) \tag{2-12a}$$

$$q_{min} = -(5/7)p \tag{2-12b}$$

and

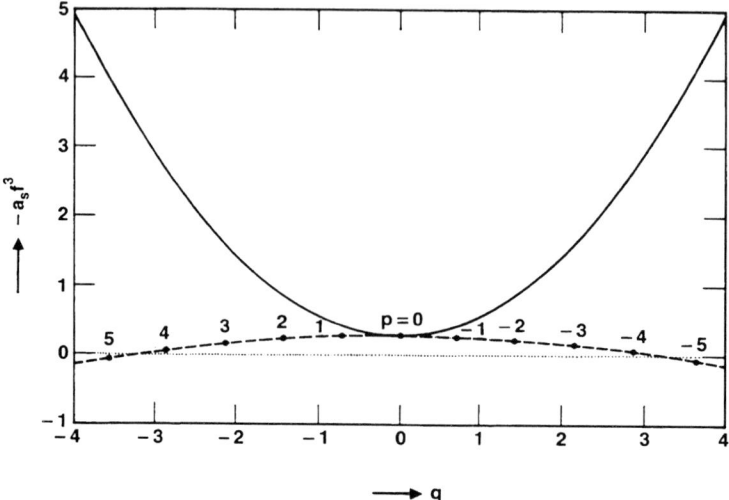

Figure 2–4. Parabolic variation of spherical aberration of a thin lens with its shape factor q for $p=0$. How its minimum value varies with q_{min} is indicated by the lower parabolic curve. Several values of p are indicated on this curve.

$$a_{smin} = -\frac{1}{32f^3}\left(9 - \frac{3}{7}p^2\right),$$

$$(2\text{-}12c)$$

respectively. Figure 2-4 shows the parabolic variation of spherical aberration with q for $p = 0$. The minimum value of spherical aberration corresponds to $q = 0$, i.e., an equiconvex lens. As pointed out earlier, the variation of spherical aberration with q for other values of p follows the same parabola except that the location of its vertex (q_{min}, a_{smin}) depends on p. The vertices of the parabolas follow the lower parabolic curve in Figure 2-4, which represents a_{smin} as a function of q_{min} obtained by substituting Eq. (2-12b) into Eq. (2-12c). The solid dots on this curve indicate various values of p. The minimum value of spherical aberration approaches zero for $|p| = \sqrt{21}$. It changes its sign for larger values of $|p|$.

It follows from Eq. (2-4b) that the coma of a thin lens is zero if its position and shape factors are related to each other according to

$$q = -\frac{(2n + 1)(n - 1)p}{n + 1} \ .$$

$$(2\text{-}13)$$

For $n = 1.5$, Eqs. (2-4b) and (2-13) reduce to

$$a_c = \frac{1}{6f^2 S'}(4p + 5q)$$

$$(2\text{-}14)$$

and

$$q = -0.8p \, , \tag{2-15}$$

respectively. For $p = -1$, the values of q giving minimum spherical aberration ($q = 0.71$) and zero coma ($q = 0.8$) are approximately equal to each other. Thus, a lens designed for zero coma for parallel incident light will have practically the minimum amount of spherical aberration. It is also possible to design and combine two thin lenses such that the spherical aberration and coma of one cancel the corresponding aberrations of the other, as illustrated by a numerical example in the next section.

2.5 Numerical Problems

As a numerical example, we determine the radii of curvature of the surfaces of a thin lens of refractive index 1.5 focusing a parallel beam of light at a distance of 15 cm from it with minimum spherical aberration. According to Eq. (2-5), $p = -1$ for a parallel beam. Substituting in Eq. (2-12b), we obtain $q_{min} = 5/7$ for minimum spherical aberration. Equation (2-6), therefore, gives

$$\frac{R_2}{R_1} = \frac{q+1}{q-1}$$
$$= -6 \, .$$

Since $f = 15$ cm, Eq. (2-1) yield $R_1 = 8.75$ cm and $R_2 = -52.50$ cm, corresponding to a nearly *plano-convex* lens with its convex side facing the incident light. For a lens of diameter 2 cm, the peak value of spherical aberration, according to Eqs. (2-3) and (2-12c), is given by $A_s = -0.79$ µm. The other primary aberrations of the focused beam can be obtained from Eqs. (2-3) and (2-4). Thus, it can be shown, for example, that the peak values of coma, astigmatism, and field curvature for a parallel beam incident on the lens at an angle of 5° from its axis, corresponding to an image height of 1.31 cm, are given by $A_c = 0.28$ µm, $A_a = -2.5$ µm, and $A_d = -2.1$ µm. A thin lens with aperture stop located at the lens does not produce any distortion. It may be noted that if the lens is turned around so that its (relatively) planar side faces the incident light, its focal length does not change. However, its shape factor changes sign, thereby changing both its spherical aberration as well as its coma.

Since spherical aberration of a thin lens varies as f^{-3}, it is possible to make it zero for a combination of lenses having focal lengths of different signs. A *doublet* designed to correct for spherical aberration can at the same time be corrected for coma. For example, we now show that two thin lenses of refractive index 1.5 focusing a parallel beam of light with radii of curvature 9.2444 cm and -15.5197 cm for the first lens, and -9.5618 cm and -15.3120 cm for the second lens, give zero spherical aberration and coma with a focal length of 15 cm when placed in contact with each other. Substituting for the refractive index and the radii of curvature of the lens surfaces into Eqs. (2-1), we find that the focal lengths of the two lenses are given by $f_1 = 11.5870$ cm and $f_2 = -50.9235$ cm. Hence, the focal length of the doublet given by $f^{-1} = f_1^{-1} + f_2^{-1}$ is $f = 15$ cm. The shape factors of the lenses are given by $q_1 = 0.2536$ and $q_2 = 4.3257$. For a parallel beam of incident light, the position factor for the first lens is given by $p_1 = -1$. Substituting for n, p_1 and q_1 into Eq. (2-12a), we find that the spherical aberration coefficient for the first lens is $a_{s1} = 2.201 \times 10^{-4}$ cm^{-3}. Since the second lens focuses the beam at a distance of $S_2' = 15$ cm, its position factor is given by $p_2 = 1 - 2f_2/S_2'$ or $p_2 = 7.7898$. Substituting for n, p_2, and q_2 into Eq. (2-12a), we find that the spherical aberration coefficient for the second lens is $a_{s2} = 2.200 \times 10^{-4}$ cm^{-3}, which is equal in magnitude but opposite in sign to the corresponding coefficient for the first lens. Hence, spherical aberration of the doublet is zero.

Now we consider the coma aberrations produced by the two lenses and the lens doublet. The first lens focuses the incident parallel beam at a distance $S_1' = f_1$. Equation (2-14) yields the coma coefficient for the first lens, $a_{c1} = 2.9281 \times 10^{-4}$ cm^{-3}. Similarly, for the second lens, $a_{c2} = 2.2618 \times 10^{-4}$ cm^{-3}. Now, for a beam incident at an angle β from the axis of the thin-lens doublet, the first lens focuses it at a height of $h_1' = \beta f_1$. The second lens forms the image of this focus at a height h_2' given by Eq. (2-2), i.e., $h_2'/h_1' = -S_2'/S_2 = S_2'/S_1' = f/f_1 = 1.2946$. If (r, θ) represent the polar coordinates of a point in the plane of the thin-lens doublet, the coma aberrations produced by the two lenses are given by

$$W_{c1}(r, \theta) = a_{c1}h_1'r^3\cos\theta$$
$$= -2.2618 \times 10^{-4}h_2'r^3\cos\theta$$

and

$$W_{c2}(r, \theta) = a_{c2}h_2'r^3\cos\theta$$
$$= -2.2618 \times 10^{-4}h_2'r^3\cos\theta \ .$$

The coma aberration of the lens doublet is given by

$$W_c(r, \theta) = W_{c1}(r, \theta) + W_{c2}(r, \theta)$$
$$= 0 \ .$$

Thus, both spherical aberration and coma of the doublet are zero.

Finally, we consider the astigmatism and field curvature aberrations of the lens doublet. Substituting for the focal length and the image distance for the two lenses in Eq. (2-4c), we obtain their astigmatism coefficients $a_{a1} = 3.2141 \times 10^{-4}$ cm^{-3} and $a_{a2} = 4.3638 \times 10^{-5}$ cm^{-3}. Hence, astigmatism aberration of the doublet at a point (r, θ) in its plane may be written

$$W_a(r, \theta; h_2') = W_{a1}(r, \theta; h_1') + W_{a2}(r, \theta; h_2')$$
$$= a_{a1}h_1'^2r^2\cos^2\theta + a_{a2}h_2'^2r^2\cos^2\theta$$
$$= (0.5967a_{a1} + a_{a2})h_2'^2r^2\cos^2\theta$$
$$= -1.4815 \times 10^{-4}h_2'^2r^2\cos^2\theta \ .$$

For a beam incident on the doublet at an angle of 5° from its axis, we obtain $h_2' = 1.31$ cm. Hence, for a beam of diameter 2 cm, the peak value of astigmatism aberration is approximately given by $A_a = 2.54\ \mu m$. Comparing Eqs. (2-4c) and (2-4d), the corresponding field curvature aberration may be obtained from A_a by multiplying it by $(n+1)/2n$. Thus, we find that $A_d = 2.12\ \mu m$.

CHAPTER 3
Aberrations of a Plane-Parallel Plate

3.1 Introduction

In Chapter 2, we considered the imaging properties of a thin lens consisting of two spherical surfaces. Now, we consider "imaging" by a *plane-parallel plate*, i.e., a plate whose two surfaces are parallel to each other, and each with a radius of curvature of infinity. Unlike a lens, such a plate is not used for imaging per se, but it is often used in imaging systems, for example, as a beam splitter or a window. The imaging relations and aberrations of a plane–parallel plate cannot be obtained from those for a thin lens in Chapter 2 by letting the radii of curvature of its surfaces approach infinity, since we neglected its thickness. However, as discussed below, they can be obtained by applying the results of Section 1.8 to its two surfaces and combining the results obtained according to the discussion of Section 1.9. It is shown that the distance between an object and its image formed by the plate, called the *image displacement*, is independent of the object position, and the aberration produced by it approaches zero as the object distance approaches infinity. Thus, a plane-parallel plate placed in the path of a converging beam not only displaces its focus by a certain amount but also introduces aberrations into it. In the case of a collimated beam, it only shifts the beam without introducing any aberrations.

3.2 Gaussian Imaging

Consider, as indicated in Figure 3-1, a plane–parallel plate of thickness t and refractive index n forming an image of a point object lying at a distance S from its front surface and at a height h from its axis. Let the aperture stop of the plate be of radius a located at its front surface.

First, we determine the location of the image formed by the plate. Using Eqs. (1-17) and (1-18) we determine the location and height of the image. For the first surface, $n = 1$, $n' = n$, and $R_1 = \infty$. Accordingly, it forms the image of P at P' such that

$$S_1' = -nS_1 \equiv -nS \tag{3-1}$$

and

$$M_1 = h_1'/h_1 = 1 , \tag{3-2}$$

where $h_1 \equiv h$. For the second surface, $n = n$, $n' = 1$, $R_2 = \infty$, and $S_2 = -S_1' + t$. Hence, it forms the image of P' at P'' such that

$$S_2' = -S_2/n = (S_1' - t)/n \tag{3-3}$$

and

$$M_2 = h_2'/h_1' = 1 . \tag{3-4}$$

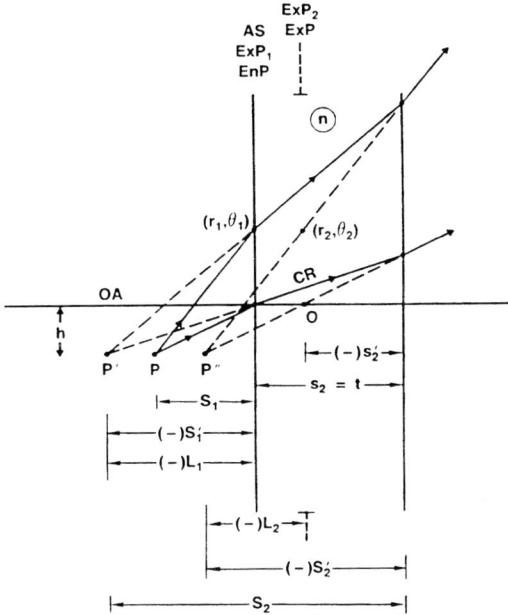

Figure 3–1. Imaging of a point object P by a plane–parallel plate of refractive index n and thickness t. P' is the image of P formed by the first surface, and P'' is the image of P' formed by the second surface of the plate. The aperture stop AS and, therefore, the entrance pupil EnP of the plate are located at the first surface. A negative sign in parentheses indicates a numerically negative quantity.

Substituting for S_1' from Eq. (3-1) into Eq. (3-3) and noting that S_2' is numerically negative, the displacement PP'' of the final image from the object may be written

$$PP'' = S_1 - (-S_2' - t)$$

$$= t(1 - 1/n) \ . \tag{3-5}$$

Thus, the image displacement is independent of the object distance S. It depends only on the thickness and the refractive index of the plate.

Next, we determine the locations and magnifications of the pupils for the two surfaces of the plate. Since the aperture stop is located at the first surface, the entrance pupil EnP of the system is also located there. Moreover, the entrance and exit pupils EnP_1 and ExP_1 for this surface are also located at the surface. The entrance pupil EnP_2 for the second surface is ExP_1. The exit pupil ExP_2 for this surface is the image of EnP_2 formed by it. Thus, letting $n = n, n' = 1, s_2 = t$, and $R_2 = \infty$, we find from Eqs. (1-17) and (1-18) that ExP_2 is located at a distance $s_2' = -t/n$ from the second surface and its magnification $m_2 = 1$. Of course, ExP_2 is also the exit pupil ExP of the system. It is evident that, for the first surface, the distance L_1 of the image P' from ExP_1 is equal to its distance S_1' from the surface. For the second surface, distance L_2 of the image P'' from ExP_2 is given by

$$L_2 = S_2' - s_2' \ , \qquad (3\text{-}6a)$$

since L_2, S_2', and s_2', are all numerically negative. Substituting for S_2' and s_2', we find that

$$L_2 = S_1' /n \qquad (3\text{-}6b)$$

$$= -S \ . \qquad (3\text{-}6c)$$

Now we use the results obtained above to determine the aberrations produced by the plate.

3.3   Primary Aberrations

First, we determine the aberration W_1 $(r_1, \theta_1; h_1')$ contributed by the first surface at a point (r_1, θ_1) in the plane of ExP_1. Letting $n = 1$, $n' = n$, and $R_1 = \infty$, Eq. (1-20) yields

$$a_{s1} = \frac{n(n^2 - 1)}{8S_1'^3} \ . \qquad (3\text{-}7)$$

Moreover, Eq. (1-22) reduces to $d_1 = -1$, and since $S_1' = L_1$, Eq. (1-21a) reduces to $a_{ss1} = a_{s1}$. The Petzval contributions to field curvature and distortion represented by the second term on the right-hand side of Eqs. (1-21d) and (1-21e) are zero. Hence, for the first surface, Eq. (1-19) may be written

$$W_1(r_1, \theta_1; h_1') = a_{s1}(r_1^4 - 4h_1' r_1^3 \cos\theta_1 + 4h_1'^2 r_1^2 \cos^2\theta_1 + 2h_1'^2 r_1^2 - 4h_1'^3 r_1 \cos\theta_1) \ . \quad (3\text{-}8)$$

Next, we determine the aberration W_2 $(r_2, \theta_2; h_2')$ contributed by the second surface at a point (r_2, θ_2) in the plane of ExP_2. Letting $n = n$, $n' = 1$, and $R_2 = \infty$, Eq. (1-20) yields for this surface

$$a_{s2} = -\frac{n^2 - 1}{8n^2 S_2'^3} \ . \qquad (3\text{-}9)$$

Once again, Eq. (1-22) reduces to $d_2 = -1$ and the Petzval contributions to field curvature and distortion are zero. Hence, for the second surface, Eq. (1-19) may be written

$$W_2(r_2, \theta_2; h_2') = a_{ss2}(r_2^4 - 4h_2' r_2^3 \cos\theta_2 + 4h_2'^2 r_2^2 \cos^2\theta_2 + 2h_2'^2 r_2^2 - 4h_2'^3 r_2 \cos\theta_2) \ , \quad (3\text{-}10)$$

where

$$a_{ss2} = (S_2' / L_2)^4 a_{s2} \ . \tag{3-11}$$

Finally, we combine the aberrations introduced by the two surfaces to obtain the aberration produced by the plate. Since m_2 and M_2 are both unity, $(r_1, \theta_1) = (r_2, \theta_2)$ and $h_2' = h_1' = h$, respectively. Hence, following Eq. (1-29), the aberration of the plane-parallel plate at a point (r, θ) in the plane of its exit pupil can be written

$$W(r, \theta; h) = W_1(r, \theta; h) + W_2(r, \theta; h) \ . \tag{3-12}$$

Substituting Eqs. (3-8) and (3-10) into Eq. (3-12), we may write the primary aberration function

$$W(r, \theta; h) = a_s (r^4 - 4hr^3 \cos\theta + 4h^2 r^2 \cos^2\theta + 2h^2 r^2 - 4h^3 r \cos\theta) \ , \tag{3-13}$$

where

$$a_s = a_{s1} + (S_2' / L_2)^4 a_{s2} \ . \tag{3-14}$$

Substituting Eqs. (3-1), (3-3), (3-6b), (3-7), and (3-9) into Eq. (3-14), it reduces to

$$a_s = \frac{(n^2 - 1) \ t}{8n^3 S^4} \ . \tag{3-15}$$

Note that the aberration increases linearly with the plate thickness t. However, as expected, it reduces to zero for a collimated incident beam ($S \to \infty$). This is indeed why a lens designer places beam splitters and windows in an imaging system in its collimated spaces wherever possible.

3.4   Numerical Problem

As a numerical example we determine the aberrations of a plane-parallel plate placed in the path of a converging beam as shown in Figure 3-2. The plate has a refractive index of 1.5. Its thickness is 1 cm and its diameter is 4 cm. In the absence of the plate, the beam comes to a focus at P at a distance of 8 cm from its front surface at a height of 0.5 cm from its axis. From Eq. (3-5), we find that the plate displaces the image from P to P' which is at the same height as P but at a distance of 8.33 cm from its front surface. Substituting for n, t, and $S = -8$ cm in Eq. (3-15), we obtain $a_s = 9.6 \times 10^{-6}$ cm^{-3}. Noting that the maximum value of r is 2 cm, we obtain the peak values of the primary aberrations introduced by the plate from Eq. (3-13); $A_s = 1.5 \ \mu$m, $A_c = -1.5 \ \mu$m, $A_a = 0.38 \ \mu$m, $A_d = 0.19 \ \mu$m, and $A_t = -0.10 \ \mu$m.

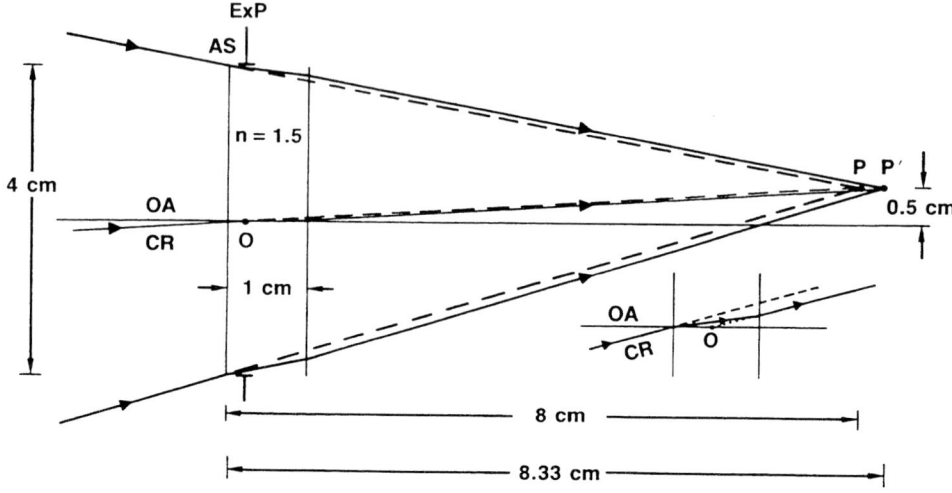

Figure 3–2. Plane-parallel plate placed in a converging beam of light. Rays incident on the plate converging toward *P* converge toward *P′* after refraction by it.

CHAPTER 4

Aberrations of a Spherical Mirror

4.1 Introduction

So far, we have considered refracting imaging systems: a spherical refracting surface in Chapter 1, a thin lens in Chapter 2, and a plane–parallel plate in Chapter 3. Now we consider the imaging properties of a spherical reflecting surface, i.e., a *spherical mirror*. These properties can be obtained in a manner similar to that for a spherical refracting surface. However, the geometry of the problem is different since now a ray incident on the surface is reflected back into the same medium containing the incident ray, instead of being refracted into another medium. Accordingly, it is instructive to draw object and image rays and not blindly use the imaging and aberration relations appropriate for a reflecting surface. In this chapter, we give the relations describing the primary aberrations of a spherical mirror for an arbitrary position of the aperture stop. These relations are applied to specific cases, one when the aperture stop is located at the mirror and the other when it is located at its center of curvature. It is shown that, in the first case, field curvature and distortion are zero. In the second case, coma, astigmatism, and distortion are zero. A numerical problem illustrates these results.

4.2 Primary Aberration Function

Consider an imaging system consisting of a spherical mirror of radius of curvature R and focal length f. Let the aperture stop and the corresponding exit pupil of the system be located as indicated in Figure 4-1. The line joining the center of curvature C of the mirror and the center of the aperture stop (and, therefore, the center O of the exit pupil) defines the optical axis of the system. Consider an object lying at a distance S from the vertex V_0 of the mirror. Let the height of an object point P from the optical axis be h. The distance S' and height h' of its Gaussian image P' are given by

$$\frac{1}{S} + \frac{1}{S'} = -\frac{2}{R} = \frac{1}{f} \tag{4-1}$$

and

$$M = \frac{h'}{h} = \frac{S' + R}{S + R} \tag{4-2a}$$

$$= -S'/S , \tag{4-2b}$$

respectively, where M is the magnification of the image.

35

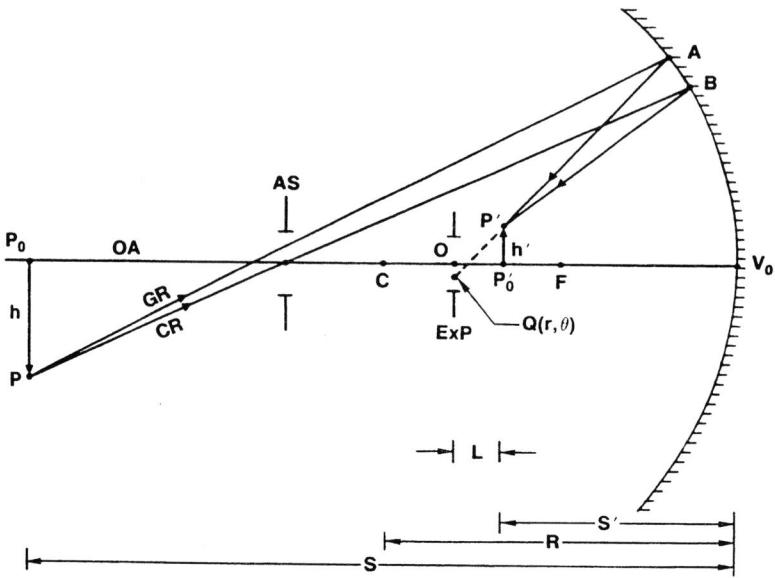

Figure 4–1. Imaging by a spherical mirror of radius of curvature R. The line joining the center of the aperture stop AS and the center of curvature C of the mirror defines the optical axis OA. The chief ray CR from a point object P passes through the center of the aperture stop.

The aberration $W(Q)$ of an object ray incident at a point A on the mirror passing through a point Q in the plane of the exit pupil with polar coordinates (r, θ) with respect to the chief ray passing through the center O of the exit pupil is given by

$$W(Q) = [PAP'] - [PBP'] .$$

It can be shown that, up to the fourth order in pupil and object or image coordinates, the aberration $W(A) \equiv W(Q)$ reduces to

$$W_s(r, \theta; h') = a_{ss}r^4 + a_{cs}h'r^3\cos\theta + a_{as}h'^2r^2\cos^2\theta + a_{ds}h'^2r^2 + a_{ts}h'^3r\cos\theta , \qquad (4\text{-}3)$$

where

$$a_s = \frac{1}{4R}\left(\frac{1}{R} + \frac{1}{S'}\right)^2 \qquad (4\text{-}4a)$$

$$= \frac{1}{4R}\left(\frac{1}{R} + \frac{1}{S}\right)^2 \qquad (4\text{-}4b)$$

$$a_{ss} = (S'/L)^4 a_s \tag{4-5a}$$

$$a_{cs} = 4d a_{ss} \tag{4-5b}$$

$$a_{as} = 4d^2 a_{ss} \tag{4-5c}$$

$$a_{ds} = 2d^2 a_{ss} - \frac{1}{2RL^2} \tag{4-5d}$$

$$a_{ts} = 4d^3 a_{ss} - \frac{d}{RL^2} \tag{4-5e}$$

$$d = -\frac{R + S' - L}{R + S'} \, , \tag{4-6}$$

and L is the distance of the Gaussian image plane from the plane of the exit pupil. For a reflecting surface, L is positive when the image lies to the left of the exit pupil. Thus, it is numerically negative in Figure 4-1. We note from the equations given above that, unless a_s is zero, coma, astigmatism, and distortion of a spherical mirror are zero when $d = 0$. As discussed in Section 4.4, this happens when the aperture stop of the mirror is located at its center of curvature. As in the case of a spherical refracting surface, spherical aberration and coma are zero when the object is located at the center of curvature of the mirror, i.e., when $S = -R$.

Comparing Eqs. (4-1)–(4-6) with Eqs. (1-17)–(1-22), we note that the results for a reflecting surface can be obtained from those for a refracting surface if we let $n = 1$ (since the mirror is in air), $n' = -1$ (minus sign represents reflection), $S' \to -S$, and $L \to -L$ (since the sign convention for image distance for reflection is the opposite of that for refraction).

4.3   Aperture Stop at the Mirror

If the aperture stop is located at the mirror as in Figure 4-2, then the entrance and exit pupils are also located there. Accordingly, $L = S'$ and $a_{ss} \to a_s$ and $d \to -R/(R + S')$. The primary aberration function given by Eq. (4-3) becomes

$$W_s(r, \theta; h') = \frac{1}{4R}\left(\frac{1}{R} + \frac{1}{S'}\right)^2 r^4 - \frac{R + S'}{R^2 S'^2} h' r^3 \cos\theta + \frac{1}{RS'^2} h'^2 r^2 \cos^2\theta \, . \tag{4-7}$$

It represents the optical path difference of a ray such as PQP' with respect to the chief ray PV_0P' in Figure 4-2 up to the fourth order in pupil and object (or image) coordinates. Note that the field curvature and distortion coefficients are zero.

If the object is located at infinity as in astronomical observations, then

$$S' = -R/2 = f \tag{4-8}$$

and

$$d = -2 \, . \tag{4-9}$$

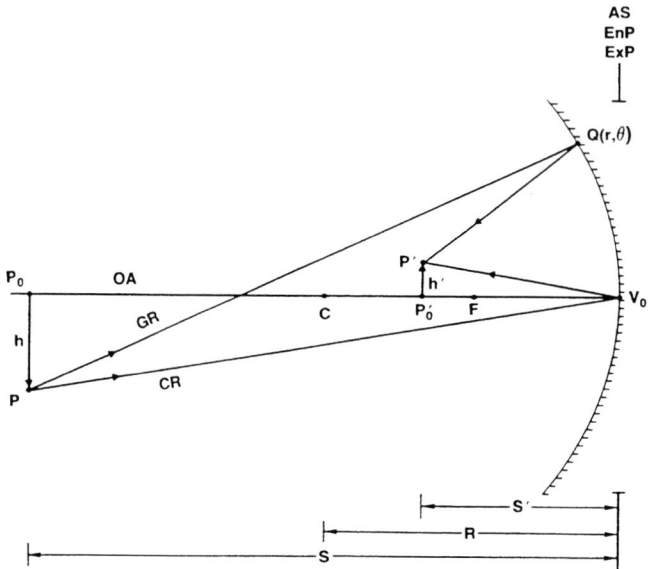

Figure 4–2. Same as Figure 4–1 except that the aperture stop is located at the mirror surface.

If it lies at an angle β from the optical axis, then

$$h' = \beta f .$$

(4-10)

Substituting Eqs. (4-8)–(4-10) into Eq. (4-7), we obtain the primary aberration function for a spherical mirror for an object at infinity at an angle β from its optical axis:

$$W_s(r, \theta; \beta) = -\frac{1}{32f^3}r^4 + \frac{1}{4f^2}\beta r^3\cos\theta - \frac{1}{2f}\beta^2 r^2\cos^2\theta .$$

(4-11)

4.4 Aperture Stop at the Center of Curvature of the Mirror

If the aperture stop is located at the center of curvature of the mirror as indicated in Figure 4-3, then the entrance pupil is also located there. The exit pupil, which is the image of the aperture stop by the mirror, is also located there, as may be seen by letting $S = -R$ (since R is numerically negative) in Eq. (4-1). The distance L of the image from the exit pupil is numerically negative since it lies to the right of the exit pupil. Accordingly,

$$L = R + S'$$

(4-12)

and, therefore, Eqs. (4-5a) and (4-6) become

$$a_{ss} = \frac{S'^2}{4R^3(R + S')^2}$$

(4-13)

and

$$d = 0 ,$$

(4-14)

respectively. Letting $d = 0$ in Eqs. (4-5b)–(4-5e), and substituting the results obtained into Eq. (4-3), we obtain the primary aberration function

$$W_s(r; h') = \frac{S'^2 r^4}{4R^3(R + S')^2} - \frac{h'^2 r^2}{2R(R + S')^2} \; . \tag{4-15}$$

Thus, coma, astigmatism, and distortion of a spherical mirror with aperture stop at its center of curvature are zero. A concave mirror has negative spherical aberration but positive field curvature. If the image is observed on a spherical surface of radius of curvature $R/2$ at a distance S' from the mirror, then the second term on the right-hand side of Eq. (4-15) representing the field curvature also vanishes. The spherical image surface is, of course, the Petzval image surface. It should be noted that, in going from Eq. (4-7) to Eq. (4-15), the maximum value of r, i.e., the radius of the exit pupil, has been multiplied by a factor of $(R + S)/S$ or $-(R + S')/S'$. Hence, the peak value of spherical aberration has not changed, as expected, owing to a change in the position of the aperture stop.

For a point object at infinity

$$S' = -R/2 \tag{4-16}$$

and, therefore,

$$L = R/2 \; , \tag{4-17}$$

and the spherical image surface of radius of curvature $R/2$ is concentric with the mirror. The spherical aberration is given by

$$a_{ss} = 1/4R^3 = -1/32f^3 \; ; \tag{4-18}$$

i.e., it is the same as for a mirror with aperture stop at its surface, as expected. It can be eliminated by placing, at the center of curvature of the mirror, a glass plate whose thickness varies as r^4. This indeed is the principle of the Schmidt system which will be discussed in Chapter 5.

It is not difficult to see why all primary aberrations, except spherical, vanish when the aperture stop is located at the center of curvature of a spherical mirror and the image is observed on the Petzval surface. Since the exit pupil is also located at the center of curvature, the chief ray corresponding to an off-axis point object passes through it. Moreover, since the mirror is spherical, any line passing through its center of curvature forms the optical axis. Hence, every point object is like an on-axis object; therefore, the only aberration that arises (with respect to its Petzval image) is spherical aberration. The Petzval curvature, corresponding to the second term on the right-hand side of Eq. (4-5d), is nonzero. It has the implication that an image aberrated by spherical aberration alone is formed on a spherical surface of radius of curvature $R/2$. This, of course, is the Petzval image surface passing through the axial image point P'_0. It is concentric with the mirror when the object is at infinity.

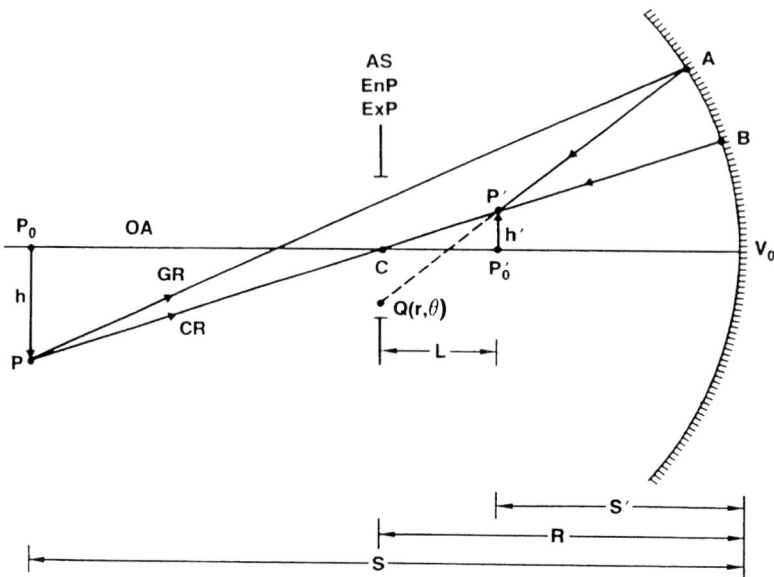

Figure 4–3. Same as Figure 4–1 except that the aperture stop is located at the center of curvature C of the mirror.

4.5 Numerical Problems

Now we consider simple numerical problems in which a spherical mirror of diameter 4 cm and a radius of curvature 10 cm images an object 2 cm high located at a distance of 15 cm from it. We assume that the aperture stop is located at the mirror and the object lies below the optical axis. Table 4-1 gives the Gaussian as well as the aberration parameters for this problem. Both concave and convex mirrors are considered in this table. The concave mirror forms a real image but the convex mirror forms a virtual image. We note that whereas astigmatism is the dominant primary aberration in the case of the concave mirror, it is coma that dominates in the case of the convex mirror. Field curvature and distortion are zero in both cases, since the aperture stop lies at the mirror surface.

Table 4-2 lists the Gaussian and aberration parameters for an object lying at infinity at an angle of 1 milliradian from the optical axis of the mirror. The magnitude of a primary aberration is independent of whether the mirror is concave or convex, but its sign depends on its type. Spherical aberration is the dominant aberration in Table 4-2. Of course, the field curvature and distortion are zero once again.

If the aperture stop of the mirror is moved to its center of curvature, the peak value A_s of its spherical aberration does not change. However, its coma and astigmatism reduce to zero, but its field curvature becomes nonzero. The radius of the exit pupil a_{ex}, the field curvature coefficient a_d, and the peak value of field curvature for the problems under consideration are given in Table 4-3. The numbers in this table without parentheses are for an object at $S = 15$ cm and those with parentheses are for an object at infinity at 1 milliradian from the optical axis of the system. As a reminder, we add that

$$a_{ex} = -aS'/(R + S')$$

$$a_d = -1/2R(R + S')^2$$

and

$$A_d = a_d h'^2 a_{ex}^2 \ ,$$

where $a = 2$ cm is the radius of the mirror. The field curvature as an aberration disappears when the image is observed on a spherical surface of radius of curvature –5 cm for the concave mirror and 5 cm for the convex mirror, located at the image plane. When the object is located at infinity, this surface is concentric with the mirror.

Table 4–1. Gaussian and aberration parameters for a spherical mirror of radius a imaging an object lying at a finite distance from it. The aperture stop is located at the mirror.

Gaussian Parameters					
Mirror	R (cm)	S' (cm)	h' (cm)	F	d
Concave	–10	7.5	1	7.5/4	–4
Convex	10	–3.75	–0.5	3.75/4	–1.6

Aberration Parameters				
Mirror	a_{ss} (cm^{-3})	A_{ss} (μm)	A_{cs} (μm)	A_{as} (μm)
Concave	-2.78×10^{-5}	–4.4	35.56	–71.1
Convex	6.94×10^{-4}	111	178	71.1

$S = 15$ cm, $h = -2$ cm, $a = 2$ cm, $S' = -RS/(R+2S)$,
$F = |S'|/2a$, $d = -R/(R+S')$

Table 4–2. Gaussian and aberration parameters for a spherical mirror imaging an object lying at infinity at an angle of 1 milliradian from its optical axis. The aperture stop is located at the mirror.

Gaussian Parameters

Mirror	R (cm)	S' (cm)	h' (cm)	F	d
Concave	−10	5	5×10^{-3}	1.25	−2
Convex	10	−5	5×10^{-3}	1.25	−2

Aberration Parameters

Mirror	a_{ss} (cm^{-3})	A_{ss} (μm)	A_{cs} (μm)	A_{as} (μm)
Concave	-2.5×10^{-4}	−40	0.8	-4×10^{-3}
Convex	-2.5×10^{-4}	40	0.8	-4×10^{-3}

Table 4–3. Radius of the exit pupil and field curvature parameters for a spherical mirror when the aperture stop is located at its center of curvature.*

Mirror	a_{ex} (cm)	a_d (cm^{-3})	A_d (μm)
Concave	2/3 (2)	8×10^{-3} (2×10^{-3})	35.6 (2×10^{-3})
Convex	10/3 (2)	-1.28×10^{-3} (-2×10^{-3})	−35.6 (-2×10^{-3})

* The numbers without parentheses are for an object at $S = 15$ cm and those with parentheses are for an object at infinity at 1 milliradian from the optical axis of the system.

CHAPTER 5

Schmidt Camera

5.1   Introduction

We have seen in Chapter 4 that a spherical mirror gives spherical aberration, which we know from Section 1.7 to be independent of the location of its aperture stop. When the aperture stop is located at the center of curvature of the mirror, it also produces field curvature, although coma, astigmatism, and distortion are all zero. As we will discuss in Chapter 6, a paraboloidal mirror forms an aberration–free image of a point object only when it lies on its axis at an infinite distance from it. In order to utilize the simplicity of fabrication of a spherical mirror, we need a way to compensate its spherical aberration. An optical system consisting of a spherical mirror and a transparent plate of nonuniform thickness placed at its center of curvature to compensate for its spherical aberration is called a *Schmidt camera*. The plate is appropriately called the *Schmidt plate*. With the exception of field curvature, the image formed is free of primary aberrations. As discussed in Section 4.4, the field curvature is such that an aberration–free image is formed on a spherical surface of radius of curvature equal to half that of the mirror. For an object at infinity, this surface is concentric with the mirror. In this chapter, we determine the shape of the Schmidt plate and discuss the chromatic aberrations associated with it. A numerical problem illustrates the results obtained.

5.2   Schmidt Plate

Consider a spherical mirror with its aperture stop located at its center of curvature C as shown in Figure 5-1 imaging an object lying at infinity. From Eq. (4-18), the optical path difference between a ray of zone r and the chief ray from an axial point object is given by

$$W(r) = -\frac{r^4}{32f^3} \ , \tag{5-1}$$

where f is the focal length of the mirror. The negative sign in Eq. (5-1) implies that the optical path length of the ray under consideration to the focus F is shorter than that of the chief ray. It also means that the ray intersects the axis after reflection at an axial point F' which is slightly closer to the mirror vertex than the paraxial focus F. This may be seen independently from the isosceles triangle CAF' in which $CF' + AF' > 2f$ or $CF' > f$ since $CF' = AF'$. In order that the path length of the ray be equal to that of the chief ray, its path length must be increased.

If a plate of refractive index n and a thickness $t(r)$ is placed at the center of curvature with a flat surface normal to the axis of the mirror, the additional optical path length introduced by the plate is given by $(n - 1) t(r)$. All object rays transmitted by the system travel equal optical path lengths and converge to a common focus F if $t(r)$ is given by

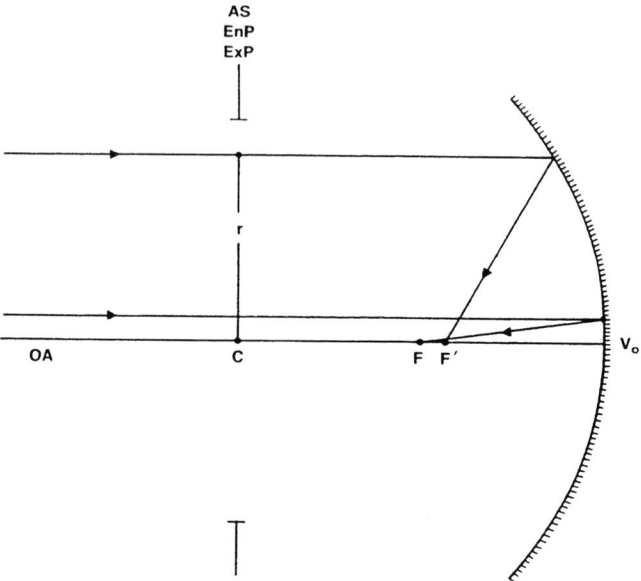

Figure 5–1. Imaging by a spherical mirror with aperture stop located at its center of curvature. Rays of different zones from an axial object at infinity intersect the axis of the mirror after reflection at different points, such as F and F', thus forming an image aberrated by spherical aberration. The ray shown intersecting the axis at F' has a zone of $\sqrt{3}a/2$, where a is the radius of the aperture stop.

$$W(r) + (n-1)t(r) = 0 . \qquad (5\text{-}2)$$

Substituting Eq. (5-1), we find that the plate thickness is given by

$$t(r) = \frac{r^4}{32(n-1)f^3} . \qquad (5\text{-}3)$$

It increases from a value of zero at its center to values proportional to the fourth power of the zonal radius. In practice, a plane-parallel plate of constant thickness t_0 would be added to it so that it can be fabricated. The shape of the plate is shown in Figure 5-2, where it is shown to slightly tilt a nonaxial ray so that, after reflection by the mirror, it passes through F.

Although spherical aberration is corrected by the use of such a plate, it does introduce *chromatic aberration*. Since the refractive index of the plate varies with the wavelength of object radiation, the angular deviation of a ray produced by the plate also varies with it. Consider a ray corresponding to a refractive index n and passing through the plate at a zone r. Since the plate is located in air and the wave aberration produced by it is $(n-1)t(r)$, following Eq. (1-1), the angular deviation of the ray produced by it is given by

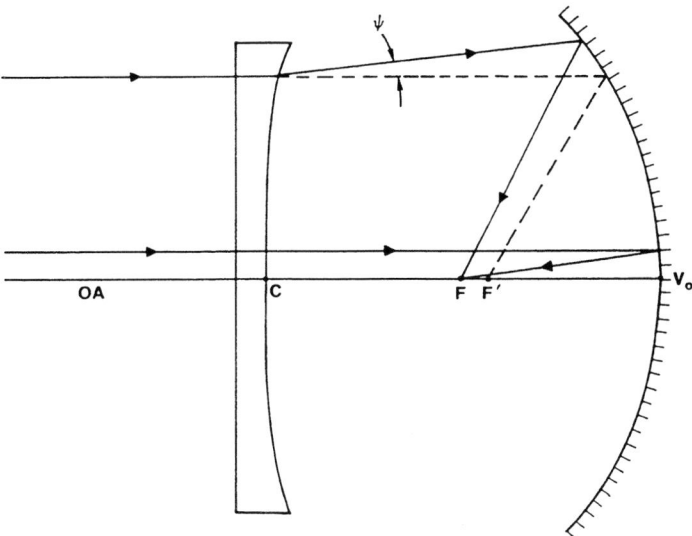

Figure 5–2. Imaging by a Schmidt camera consisting of a spherical mirror and a transparent plate placed at its center of curvature *C*. The spherical aberration of the mirror is precorrected by the plate so that the system forms an image free of spherical aberration. The dashed lines indicate the path of a ray in the absence of a Schmidt plate.

$$\psi = (n - 1)\frac{dt}{dr} \ . \tag{5-4}$$

Substituting Eq. (5-3) into Eq. (5-4), we obtain

$$\psi = \frac{r^3}{8f^3} \ . \tag{5-5}$$

From Eq. (5-4), the *angular dispersion* of any ray is given by

$$\Delta\psi = \Delta n \frac{dt}{dr} \ , \tag{5-6a}$$

where Δn is the variation in the refractive index of the plate across the spectral bandwidth of the object radiation. Substituting for dt/dr from Eq. (5-4) into Eq. (5-6a), we obtain

$$\Delta\psi = \frac{\Delta n}{n - 1}\psi \ . \tag{5-6b}$$

Thus, the angular dispersion $\Delta\psi$ of a ray produced by the plate is proportional to its angular deviation ψ. The value of ψ is maximum and equal to $a^3/8f^3$ for the marginal rays, i.e., for $r = a$, where a is the radius of the plate.

To reduce the chromatic aberration, we must reduce the maximum value of ψ. To do so, we add to the plate a very thin planoconvex lens. Such a lens will reduce the focus distance such that the rays are now focused at a point F' instead of F as in Figure

5-3. A plano-convex lens introduces thickness to the plate varying as r^2. Thus, the plate thickness may be written

$$t(r) = t_0 + \frac{r^4}{32(n-1)f^3} + \frac{br^2}{n-1} \ , \tag{5-7}$$

where b is a constant chosen to minimize the chromatic aberration. Comparing the defocus aberration br^2 introduced by the plate with Eq. (1-3b), we find that the distance between F and F' is given by $2bf^2$. F' lies on the right-hand side of F, as in Figure 5-3, if b is numerically negative. The thickness variations of plates with different values of b are shown in Figure 5-4. We note that the depth of *material removal*, starting with a plane-parallel plate, is minimum when $b = -a^2/32f^3$ (corresponding to $c = 1$ in the figure). However, we are interested in minimizing the maximum value of the angular deviation of a ray. As shown below, this requires that b be equal to $-3a^2/64f^3$ (or $c = 1.5$).

Substituting Eq. (5-7) into Eq. (5-4), we find that the angular deviation of a ray is now given by

$$\psi = \frac{r^3}{8f^3} + 2br \ . \tag{5-8}$$

Its maximum value in the range $0 \leq r \leq a$ occurs either at its stationary point $r = (-16f^3b/3)^{1/2}$ obtained by letting $\partial\psi/\partial r = 0$, or at $r = a$. At the former its absolute value is $(4|b|/3)(-16f^3b/3)^{1/2}$ and at the latter it is $(a^3/8f^3) + 2ba$. These two values are both equal to $a^3/32f^3$ if $b = -3a^2/64f^3$ (or $c = 1.5$), thus reducing the angular deviation as well as the chromatic aberration by a factor of 4 compared to their values if $b = 0$.

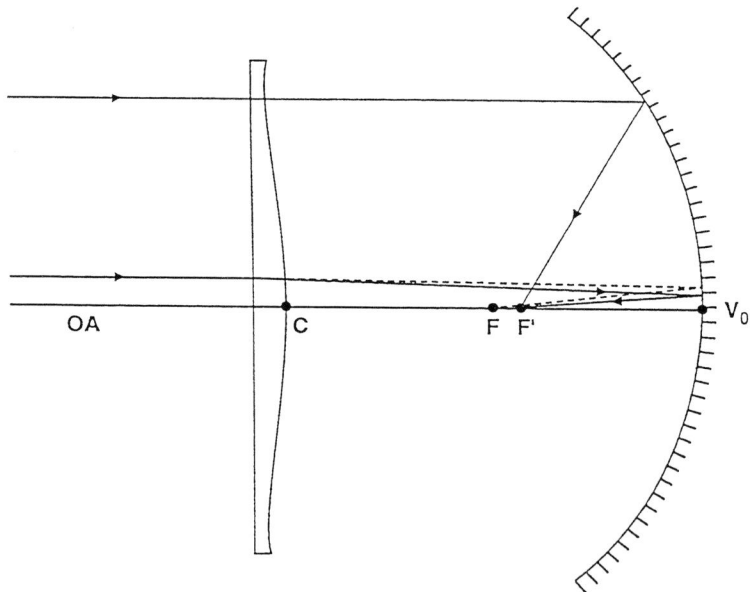

Figure 5–3. Schmidt camera with a plate introducing minimum chromatic aberration. The dashed lines indicate the path of a ray in the absence of the Schmidt plate. All rays passing through the plate and reflected by the mirror are focused at F', where the ray passing through the neutral zone of the plate is focused by the mirror.

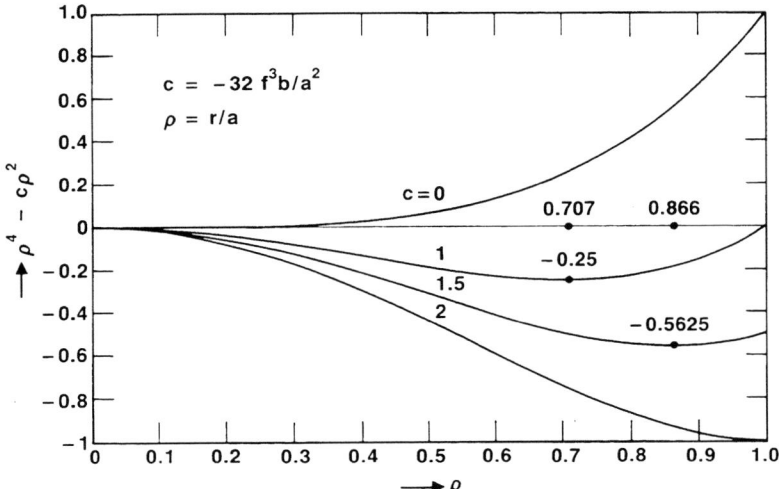

Figure 5–4. Thickness variation of a Schmidt plate for different values of b. The variation is minimum when $b = -a^2/32f^3$, corresponding to $c = 1$.

Substituting this value of b into Eq. (5-7) we find that the plate thickness required for eliminating spherical aberration introduced by the mirror and minimizing the chromatic aberration introduced by the plate is given by

$$t(r) = t_0 + \frac{1}{32(n-1)f^3}\left(r^4 - \frac{3}{2}a^2r^2\right) . \tag{5-9}$$

We note that $\partial\psi/\partial r = 0$ for $r = \sqrt{3}\,a/2$. This value of r is called the *neutral zone* of the plate since a ray incident normal to it passes through it undeviated as in Figure 5-3. As may be seen from Figure 5-4, the thickness variation of the plate and the material removal are maximum at this zone. This variation is more than twice the variation for a minimum-thickness-variation plate; compare the numbers -0.5625 and -0.25 in the figure which occur at zones of $r = 0.707a$ and $r = 0.866a$, respectively.

The lens component of the Schmidt plate has a focal length of $f_l = 32f^3/3a^2$. The vertex radius of curvature of the plate is equal to $(1-n)f_l$. This, of course, is the radius of curvature of the second surface of the lens.

The angular dispersion of the rays is now given by

$$\Delta\psi = \frac{\Delta n}{8(n-1)f^3}\left(r^3 - \frac{3}{4}a^2r\right) . \tag{5-10}$$

Its maximum value occurs for rays with $r = a/2$ and a. It is given by

$$(\Delta\psi)_{max} = \frac{\Delta n}{32(n-1)}\,\frac{a^3}{f^3} . \tag{5-11}$$

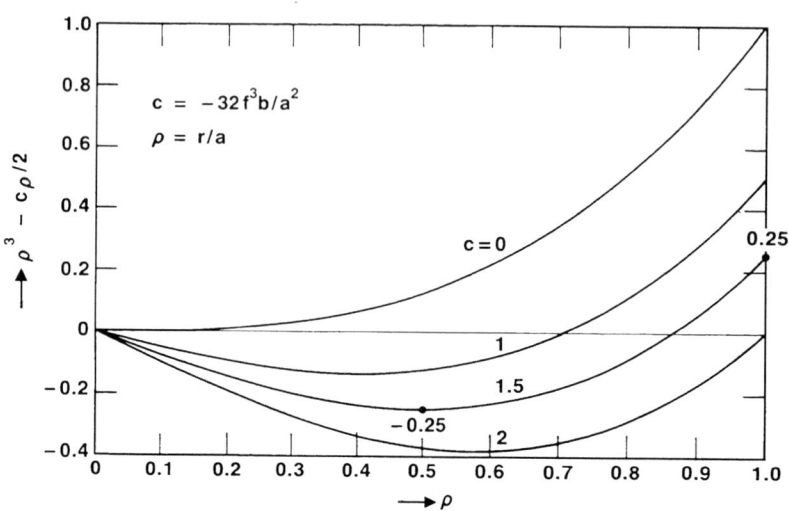

Figure 5–5. Dependence of angular dispersion on the value of b. It is minimum when $b = -3a^2/64f^3$, corresponding to $c = 1.5$.

The dependence of angular dispersion on the value of b is illustrated in Figure 5-5. We shall see in Chapter 7 that the value of b giving minimum chromatic aberration also gives the position of the defocused image plane in which the rays forming an image of a point object aberrated by spherical aberration have a minimum spot radius (circle of least confusion).

It has been shown in Section 4.4 that a spherical mirror with an aperture stop located at its center of curvature gives only spherical aberration and field curvature. The Schmidt plate compensates for the spherical aberration and, therefore, the image of an extended object observed on a spherical surface concentric with the mirror is free of primary aberrations. Strictly speaking, the lens component of the plate also introduces small amounts of primary aberrations. The spherical aberration contributed by it can be made zero by slightly adjusting the value of r^4 term in the plate thickness $t(r)$.

For an on-axis point object, the mirror contributes some *secondary* or sixth-order spherical aberration. It can be made zero by introducing an r^6 term in the plate. For off-axis point objects, the Schmidt plate given by Eq. (5-9) also introduces sixth-order aberrations. One of these is *oblique or lateral spherical aberration* varying as $\beta^2 r^4$, where β is the field angle of a point object. The other is a *sixth-order astigmatism* (called wings by Schwarzschild) varying as $\beta^2 r^4 \cos^2\theta$. The ratio of the peak values of these two aberrations is 1:4n (see Linfoot p. 190). As a result, the geometrical spot diagrams (discussed for primary aberrations in Chapter 7) for off-axis point objects have a much larger width in the tangential direction than that in the sagittal direction.

It should be noted that as the field angle β increases, the size of the focal surface also increases, which, in turn, obscures the ray bundle incident on the mirror. For a *field of view* of radius β, the linear obscuration of the on-axis beam incident on the mirror is given by $\epsilon = 2\beta F$, where F is the focal ratio of the system.

5.3   Numerical Problems

As a numerical example, we consider a spherical mirror with a radius $a = 5$ cm and a focal length $f = 40$ cm so that $F = 4$. According to Eq. (5-1), the peak value of spherical aberration introduced by it for an object at infinity is equal to 3.05 μm. If a Schmidt plate of refractive index $n = 1.5$ is used to compensate for this spherical aberration, the difference in its maximum and minimum thickness is 6.10 μm according to Eq. (5-3). Thus, starting with a plate of uniform thickness, as much as 6.1 μm deep material must be removed at its center, reducing to a value of zero at its edge. This would be satisfactory for operation in monochromatic light for which the refractive index is 1.5. The image is formed at a distance of 40 cm from the mirror. The image of an extended object lying at infinity is free of primary aberrations when observed on a spherical surface of radius of curvature 40 cm concentric with the mirror passing through its focal point F.

For white-light operation, the thickness variation of the plate for minimum chromatic aberration is given by Eq. (5-9). Thus, the plate has a certain thickness at the center and its variation is maximum and equal to 3.43 μm at its neutral zone of $\sqrt{3}\,a/2$ = 4.33 cm. Its variation at its edge is 3.05 μm. We note that the depth of material removal is less for this plate than that for the monochromatic operation. The image is now formed at a distance which is 0.586 mm closer to the mirror than its focal plane. If $\Delta n = 0.025$ across the spectral bandwidth of object radiation, then, according to Eq. (5-11), the minimum radius of the chromatic image will be 1.22 μm. In practice, the image will be larger than this due to diffraction by, say, the diameter of the Airy disc (discussed in Chapter 8). For visible light, the diameter of the Airy disc (neglecting the effect of obscuration due to the focal surface) is approximately 6.83 μm, where we have used a visible wavelength of 0.7 μm.

CHAPTER 6
Aberrations of a Conic Surface

6.1 Introduction

So far, we have considered the aberrations of spherical surfaces, which are conic surfaces of zero eccentricity. In this chapter, we discuss the aberrations of a *conic surface* with an arbitrary value of *eccentricity*. Our starting point is imaging by and aberrations of a spherical surface discussed in Sections 1.8 and 4.2. It should be noted that the Gaussian imaging equations for a conic surface of a certain *vertex radius of curvature* are the same as those for a spherical surface with the same radius of curvature. Given the aberrations of a spherical surface, we determine the additional aberrations introduced by a corresponding conic surface. In particular, we show that if the aperture stop is located at the conic surface, the only additional aberration is spherical aberration. The other (primary) aberrations of the conic surface are identically the same as those of the spherical surface. The aberrations of a conic surface are further generalized to obtain the aberrations of a *general aspherical* (nonconic) *surface*. The aberrations of a *paraboloidal mirror* are briefly discussed and compared with those of a spherical mirror. Finally, we outline a procedure to determine the aberrations of a multimirror system.

6.2 Conic Surface

A conic surface of eccentricity e and vertex radius of curvature R is described by its *sag* according to

$$z_c = \frac{r_c^2/R}{1 + [1 - (1 - e^2)\, r_c^2/R^2]^{1/2}}, \tag{6-1}$$

where, as illustrated in Figure 6-1, (x_c, y_c, z_c) are the coordinates of a point on it and

$$r_c = (x_c^2 + y_c^2)^{1/2}, \tag{6-2}$$

is the distance of the point from the z axis. The origin of the coordinate system is at the vertex of the conic, and the z axis is along its axis of rotational symmetry. The various conic surfaces are described by their values of e according to

$$
\begin{array}{lll}
e & = 1 & \text{Paraboloid} \\
 & < 1 & \text{Ellipsoid} \\
 & > 1 & \text{Hyperboloid} \\
 & = 0 & \text{Sphere.}
\end{array}
$$

If we neglect terms in r_c of order higher than four, Eq. (6-1) becomes

$$z_c \simeq \frac{r_c^2}{2R} + (1 - e^2)\frac{r_c^4}{8R^3}. \tag{6-3}$$

50

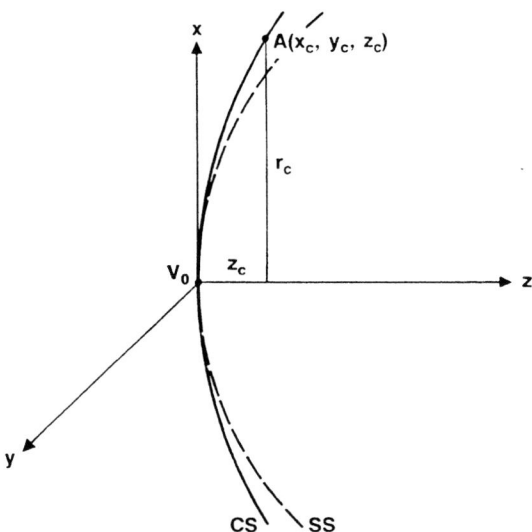

Figure 6–1. Sag of a conic surface. The origin of the coordinate system lies at the vertex V_0 of the conic. The axis about which the conic is rotationally symmetric is the z axis of the coordinate system. z_c is the sag of a point A on the conic.

Thus, up to the fourth order in r_c, the sag of a spherical ($e = 0$) surface is larger than that of a conic surface by $e^2 r_c^4 / 8R^3$. Up to this order, the chord $V_0 A \simeq r_c$.

6.3   Conic Refracting Surface

6.3.1 On-Axis Point Object

Consider a conic surface separating media of refractive indices n and n'. Compared to a spherical surface, a conic surface introduces an *additional* aberration for an axial point object P_0, which for a ray passing through a point A on the spherical surface in Figure 6-2 is given by

$$\Delta W_c (A) \simeq (n' - n)\overline{A}_0 A , \qquad (6\text{-}4a)$$

where

$$\overline{A}_0 A \simeq e^2 r_c^4 / 8R^3 , \qquad (6\text{-}4b)$$

is approximately equal to the sag difference between a sphere and a conic of the same vertex radius of curvature. Since $r_c \simeq V_0 A$, we may write

$$\Delta W_c (A) = \sigma V_0 A^4 , \qquad (6\text{-}5a)$$

where

$$\sigma = (n' - n) \, e^2 / 8R^3 . \qquad (6\text{-}5b)$$

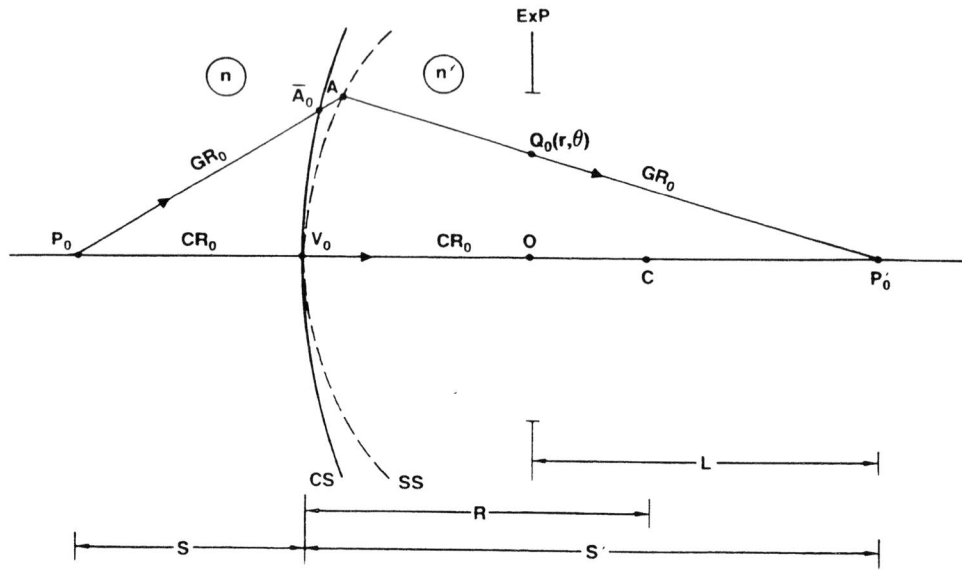

Figure 6–2. Imaging of an on-axis point object P_0 by a conic refracting surface CS of vertex radius of curvature R and center of curvature C. The Gaussian image is located at P_0'.

We note from (approximate) triangles $V_0 AP'$ and $OQ_0 P_0'$ in the figure that $V_0 A/OQ_0 = S'/L$. Hence, the aberration at a point Q_0 in the plane of the exit pupil at a distance r from the optical axis is given by

$$\Delta W_c(Q_0) = \sigma(S'/L)^4 OQ_0^4$$

or

$$\Delta W_c(r) = \sigma(S'/L)^4 r^4. \tag{6-6}$$

6.3.2 Off-Axis Point Object

For an off-axis point object such as P in Figure 6-3, the optical path length of the chief ray for a conic surface is also different from that for a spherical surface. Accordingly, the conic contribution to the aberration of a ray from the point object P and passing through a point A on the spherical surface is given by

$$\Delta W_c(A) \simeq (n'-n)\,(\overline{A}A - \overline{B}B)$$

$$= \sigma(V_0 A^4 - V_0 B^4). \tag{6-7}$$

Let (r,θ) be the polar coordinates of a point Q, where the ray under consideration intersects the plane of the exit pupil, with respect to O as the origin. From Figure 6-4, which represents the projection of the exit pupil on the refracting surface, we note that

$$V_0 A^2 = AB^2 + V_0 B^2 - 2AB\ V_0 B\cos\theta\ . \tag{6-8}$$

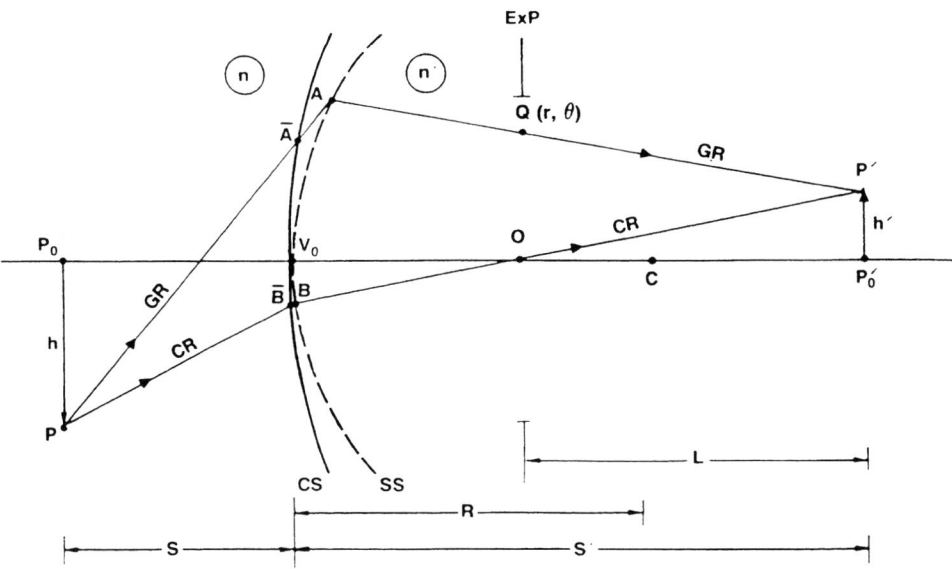

Figure 6–3. Imaging of an off-axis point object P by a conic refracting surface of vertex radius of curvature R and center of curvature C. The Gaussian image is located at P'.

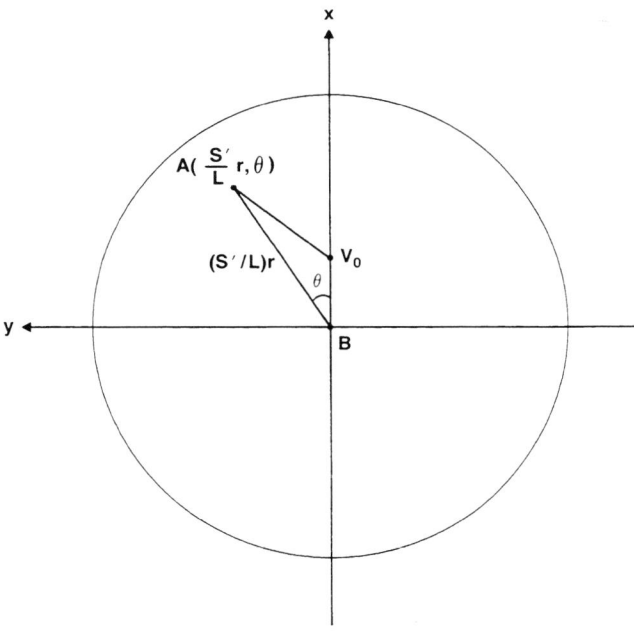

Figure 6–4. Projection of the exit pupil on the refracting surface in Figure 6–3. Point B, which lies on the chief ray, forms the center of the projected pupil.

Also, from (approximate) triangles BAP' and OQP' in Figure 6-3, we note that

$$AB \simeq (S'/L)\, r \ . \tag{6-9a}$$

Similarly, from (approximate) triangles $OV_0 B$ and $OP_0' P'$,

$$V_0 B \simeq gh' \ , \tag{6-9b}$$

where

$$g = \frac{S' - L}{L} \ . \tag{6-10}$$

Substituting Eqs. (6-9) into Eq. (6-8), squaring the result and then substituting into Eq. (6-7), we obtain

$$\Delta W_c(Q) = \sigma\,[(S'/L)^4 r^4 - 4(S'/L)^3 gh'r^3\cos\theta$$

$$+ 4(S'/L)^2 g^2 h'^2 r^2\cos^2\theta + 2(S'/L)^2 g^2 h'^2 r^2$$

$$- 4(S'/L)g^3 h'^3 r\cos\theta\,] \ . \tag{6-11}$$

Adding Eqs. (6-11) and (1-19), we obtain the primary aberrations of a conic surface. Note that if the aperture stop is located at the conic surface so that $S' = L$ and, in turn, $g = 0$, then its aberrations differ from those of a spherical surface only in spherical aberration. The other primary aberrations are identical for the two surfaces.

6.4 General Aspherical Refracting Surface

Consider a *general* rotationally symmetric *aspherical surface* with a vertex radius of curvature R and described by its sag according to

$$z_g = \frac{r_g^2}{2R} + (1 - e^2 + s_g)\frac{r_g^4}{8R^3} \ , \tag{6-12}$$

where s_g represents the fourth-order sag contribution over and above that of a conic surface of the same vertex radius of curvature. Compared to a spherical surface of radius of curvature R, the general aspherical surface contributes an additional optical path length to a zonal ray from an axial point object given by

$$\Delta W_g(A) = \sigma' V_0 A^4 \ , \tag{6-13a}$$

where

$$\sigma' = (n' - n)\,(e^2 - s_g)/8R^3 \ . \tag{6-13b}$$

Comparing Eqs. (6-5) and (6-13), we note that if we replace σ by σ' in the results obtained for a conic surface, we obtain the aberrations for a general aspherical surface. This conclusion applies to the aberrations for off-axis point objects as well.

6.5 Conic Reflecting Surface

The additional primary aberrations introduced by a conic or a general aspherical *reflecting* surface, compared to a spherical one (discussed in Chapter 4), can be obtained from those for a corresponding refracting surface by letting $n = -n' = 1, S' \rightarrow$

$-S'$, and $L \rightarrow -L$ (since the sign convention for image distance for reflection is opposite to that for refraction). Thus, for example, the additional aberration of a *conic mirror* is given by Eq. (6-11), where

$$\sigma = -e^2/4R^3 \ . \tag{6-14}$$

6.6 Paraboloidal Mirror

For a paraboloidal ($e = 1$) mirror, we note from Eqs. (4–4) and (6–14) that for an object at infinity

$$a_s = -\sigma \tag{6-15a}$$

$$= 1/4R^3 \ . \tag{6-15b}$$

Hence, following Eqs. (4-3) and (4-11), we find that the spherical aberration of a paraboloidal mirror is zero when the object lies at infinity, i.e.,

$$a_{sc} = (S'/L)^4 \ (a_s + \sigma)$$

$$= 0 \ . \tag{6-16}$$

If, in addition, the aperture stop is located at the mirror, then $S' = L$ and, therefore, $g = 0$. Hence, Eq. (6-11) shows that the other primary aberrations of a paraboloidal mirror are identical with those for a spherical mirror. Accordingly, the image of an off-axis object at infinity formed by a paraboloidal mirror with stop at the mirror surface suffers only from coma and astigmatism given by the corresponding terms in Eq. (4-11). Thus, for example, the image of an object lying at infinity at an angle of 1 milliradian from the axis of a paraboloidal mirror of vertex radius of curvature of 10 cm suffers from coma aberration with a peak value of 0.8 μm but a negligible value of astigmatism. These values are the same as those of the corresponding aberrations in Table 4-2 where imaging by a spherical mirror was considered. Thus, the difference between imaging by paraboloidal and spherical mirrors lies in their spherical aberrations: zero in the case of a paraboloidal mirror and – 40 μm peak aberration, for example, in the case of a concave spherical mirror. Of course, the image of an axial object at infinity by a paraboloidal mirror is aberration free.

6.7 Multimirror Systems

The aberrations of a multimirror system can be calculated by determining the aberrations of each mirror at its respective exit pupil and then combining them according to the procedure described in Section 1.9. Thus, for example, we can show that an afocal system consisting of two confocal paraboloidal mirrors acting as a beam expander is anastigmatic, introducing only field curvature and distortion aberrations. Similarly, we can investigate the aberrations of two–mirror systems such as the *Cassegrain* and *Ritchey-Chrétien* telescopes. As is sometimes the case in practice, this is easier said than done.

CHAPTER 7

Ray Spot Diagrams

7.1   Introduction

In Chapters 2–6, we have determined the primary wave aberrations of simple optical imaging systems. In this chapter, we use the relationship between the wave and ray aberrations given in Section 1.2 to determine the ray distribution for a point object, called the *ray spot diagrams*, in the Gaussian image plane. For each primary aberration, we determine the extent or the size of the image spot in terms of its peak value and the focal ratio of the image forming light cone. In the case of spherical aberration and astigmatism, we consider ray distributions in defocused image planes as well and determine the plane in which the spot size is minimum. These minimum–size spots are referred to as the *circles of least confusion* and represent the best aberrated images based on geometrical optics. In lens design, one often tries to minimize the radius of gyration of the spot rather than its radius. However, we will see in Chapter 8 that, in reality, which is based on diffraction of light at the exit pupil of the system, an image distribution is not given by the corresponding ray spot diagrams. For example, the aberration-free image of a point object is a point according to geometrical optics, but its diffraction image for a circular pupil consists of a bright spot surrounded by concentric dark and bright rings. However, certain features of the aberrated images of the two types are common.

7.2   Wave and Ray Aberrations

Consider an optical system consisting of a series of rotationally symmetric coaxial refracting and/or reflecting surfaces imaging a point object. We have discussed in Chapter 1 that the primary aberration function representing the *wave aberration* at its exit pupil can be written

$$W(r, \theta; h') = a_s r^4 + a_c h' r^3 \cos\theta + a_a h'^2 r^2 \cos^2\theta + a_d h'^2 r^2 + a_t h'^3 r \cos\theta , \quad (7\text{-}1)$$

where (r, θ) are the polar coordinates of a point in the plane of the exit pupil, h' is the height of the Gaussian image point, and a_s, a_c, a_a, a_d, and a_t represent the coefficients of *spherical aberration, coma, astigmatism, field curvature,* and *distortion,* respectively. The angle θ is equal to zero or π for points lying in the *tangential* or *meridional plane* (i.e., the plane containing the optical axis and the point object and, therefore, its Gaussian image). The *chief ray*, which, by definition, passes through the center of the exit pupil, always lies in this plane. The plane normal to the tangential plane but containing the chief ray is called the *sagittal plane*. As the chief ray bends when it is refracted or reflected by a surface, so does the sagittal plane.

For an optical system with a circular exit pupil, say, of radius a, it is convenient to use normalized coordinates (ρ, θ) where $\rho = r/a, 0 \le \rho \le 1, 0 \le \theta < 2\pi$, suppress the explicit dependence on h', and write the aberration function in the form

$$W(\rho, \theta) = A_s \rho^4 + A_c \rho^3 \cos\theta + A_a \rho^2 \cos^2\theta + A_d \rho^2 + A_t \rho\cos\theta \ , \qquad (7\text{-}2)$$

where the new aberration coefficients A_i, representing the peak or maximum values of the aberrations, are related to those used in Eq. (7-1) according to

$$A_s = a_s a^4, \ A_c = a_c h'a^3, \ A_a = a_a h'^2 a^2, \ A_d = a_d h'^2 a^2, \ A_t = a_t h'^3 a \ . \qquad (7\text{-}3)$$

Although we will discuss the spot diagrams in terms of these *peak aberration coefficients*, it is necessary to know their dependence on the image height h' when discussing images of extended objects.

If (x,y) represent the rectangular coordinates of a pupil point, the corresponding normalized coordinates (ξ, η) are given by

$$(\xi, \eta) = \frac{1}{a}(x,y) \qquad (7\text{-}4a)$$

$$= \rho(\cos\theta, \sin\theta) \ . \qquad (7\text{-}4b)$$

The aberration function defined in the form of Eq. (7-2) has the advantage that the aberration coefficients A_i have the dimensions of length (i.e., dimensions of the wave aberration), and they represent the peak or maximum values of the corresponding primary aberrations. For example, if $A_s = 1\lambda$, where λ is the wavelength of the object radiation, we speak of one wave of spherical aberration.

The distribution of rays in an image plane is called the *ray spot diagram*. Their density (i.e., the number of rays per unit area) is called the *geometrical point-spread function* (PSF). If the system is aberration free, then the wavefront is spherical and all the object rays transmitted by the system converge to the Gaussian image point. When the wavefront is aberrated, a ray passing through a point (ρ, θ) in the plane of the exit pupil intersects the Gaussian image plane at a point which, following Eq. (1-1), may be written

$$(x_i, y_i) = 2F\left(\frac{\partial W}{\partial \xi}, \frac{\partial W}{\partial \eta}\right) \qquad (7\text{-}5a)$$

$$= 2F\left(\cos\theta \frac{\partial W}{\partial \rho} - \frac{\sin\theta}{\rho}\frac{\partial W}{\partial \theta} \ , \sin\theta \frac{\partial W}{\partial \rho} + \frac{\cos\theta}{\rho}\frac{\partial W}{\partial \theta}\right) \ , \qquad (7\text{-}5b)$$

where $F = R/2a$ is the focal ratio or the f–number of the image-forming light cone. Here, (x_i, y_i) represent the *ray aberrations*, i.e., the coordinates of the point of intersection of the ray in the Gaussian image plane with respect to the Gaussian image point and R is the radius of curvature of the *Gaussian reference sphere* with respect to which the aberration $W(\rho, \theta)$ is defined. The reference sphere is centered at the Gaussian image point $(0,0)$ and, like the aberrated wavefront, passes through the center of the exit pupil. In Eqs. (7-5), we have assumed that the refractive index of the medium in which the image is formed is unity since it is often the case in practice. Substituting Eq. (7-2) into Eq. (7-5b), we find that, in the absence of distortion, the chief ray intersects the Gaussian image plane at the Gaussian image point.

For a radially symmetric aberration, i.e., one for which $W(\rho,\theta) = W(\rho)$, we note from Eq. (7-5b) that the PSF is also radially symmetric. The radial distance r_i of a ray from the Gaussian image point in that case is given by

$$r_i = \left(x_i^2 + y_i^2\right)^{1/2}$$
$$= 2F\frac{\partial W(\rho)}{\partial \rho} \ . \tag{7-6}$$

Now we discuss the characteristics of an image aberrated by a primary aberration. To be definite, we assume that each of the aberration coefficients A_i is positive, unless stated otherwise.

7.3   Spherical Aberration

Figure 7-1 illustrates the relationship between a wavefront aberrated by spherical aberration

$$W(\rho) = A_s\rho^4 \ , \tag{7-7}$$

and the reference sphere centered at a Gaussian image point P_0'. Substituting Eq. (7-7) into Eq. (7-6) we find that a ray of *zone* ρ intersects the Gaussian image plane at a distance

$$r_i = 8FA_s\rho^3 \ , \tag{7-8}$$

from P_0'. Thus, the rays lying on a circle of radius ρ in the exit pupil lie on a circle of radius r_i given by Eq. (7-8) in the Gaussian image plane. The maximum value of r_i is $8FA_s$ and corresponds to rays with $\rho = 1$; i.e., it corresponds to the marginal rays. We shall refer to the maximum value of r_i as the radius of the image spot. Note that since A_s is independent of the height h of the point object from the optical axis, the ray distribution owing to spherical aberration alone is also independent of h.

Let us consider the ray distribution in a slightly defocused image plane by introducing a defocus aberration A_d. The aberration with respect to a reference sphere centered at a defocused point may be written

$$W(\rho) = A_s\rho^4 + A_d\rho^2 \ . \tag{7-9}$$

The rays of zone ρ now lie in the defocused image plane on a circle of radius

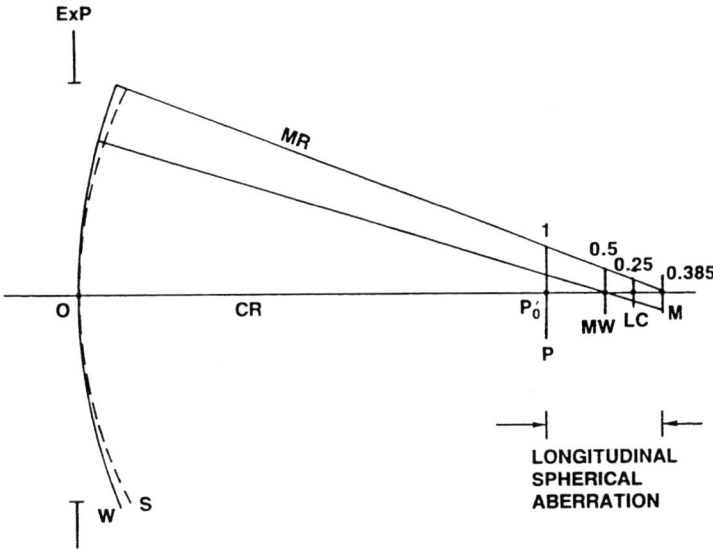

Figure 7-1. Ray spot radii in various image planes for a wavefront W aberrated by spherical aberration. P – paraxial, M – marginal, MW – midway, LC – least confusion. The reference sphere S is centered at the Gaussian image point P_0'. The values of spot radii indicated in the figure are in units of $8FA_s$, where F is the focal ratio of the image-forming light cone and A_s is the peak value of the spherical aberration.

$$r_i = 8FA_s[\rho^3 + (A_d/2A_s)\rho] . \tag{7-10}$$

For the *marginal rays*, corresponding to $\rho = 1$, $r_i \to 0$ if $A_d = -2A_s$. Following Eqs. (1-3c) and (1-3d), we find that the marginal rays intersect the axis at a distance

$$\Delta R = 8F^2A_d \tag{7-11}$$

$$= -16F^2A_s , \tag{7-12}$$

from P_0'. This distance shown as $P_0' M$ in Figure 7-1 is called the *longitudinal spherical aberration*. A negative value of ΔR implies that, compared to the old reference sphere, the new reference sphere is centered at a point which is farther from the center of the exit pupil. Hence, the point of intersection M of the marginal rays lies to the right of P_0' as shown in Figure 7-1. This is to be expected for positive values of A_s. The points P_0' and M are called the *paraxial* (meaning for very small values of ρ) and the *marginal image points*. Substituting $A_d = -2A_s$ in Eq. (7-10), we find that the maximum value of r_i in the *marginal image plane* occurs for rays of zone $\rho = 1/\sqrt{3}$. This maximum value is $2/3\sqrt{3}$ (or 0.385) times the corresponding value in the *paraxial* (Gaussian) *image plane*. Thus, the *marginal spot radius* is considerably smaller than the *paraxial spot radius*.

The image plane lying midway between the paraxial and marginal planes corresponds to $A_d = -A_s$. The spot radius in this plane is half of that in the paraxial plane and corresponds to marginal rays. Comparing Eq. (7-10) with Eq. (5-8), we find that the spot radius is minimum in a plane corresponding to $A_d = -3A_s/2$, i.e., a plane which is 3/4 of the way from the paraxial plane to the marginal plane. The spot radius in this case is 1/4 of the paraxial spot radius and corresponds to rays of zone $\rho = 1/2$ and 1. This spot is called the *circle of least (spherical) confusion*. The spot radii in various image planes are listed in Table 7-1.

The mixing of one aberration with one or more aberrations is called *aberration balancing*. Here, we have balanced spherical aberration with defocus in order to reduce the spot size. The amount of defocus that gives the smallest ray spot is called the optimum defocus based on geometrical optics. We shall see in Chapter 8 that, based on diffraction, the optimum amount of defocus is different, since there it is used to reduce the variance of the aberration across the exit pupil.

Table 7–1. Ray spot radii in units of $8FA_s$ for spherical aberration A_s

Image Plane	Balancing Defocus A_d/A_s	Spot Radius
Paraxial	0	1
Marginal	−2	0.385
Midway	−1	0.5
Least Confusion	−3/2	0.25

7.4 Coma

The coma wave aberration is given by

$$W(\rho, \theta) = A_c \rho^3 \cos\theta = A_c \xi (\xi^2 + \eta^2) .\qquad (7\text{-}13)$$

Substituting Eq. (7-13) into Eq. (7-5), we obtain the corresponding ray aberrations in the Gaussian image plane with respect to the Gaussian image point. They are given by

$$(x_i, y_i) = 2FA_c\rho^2 (2 + \cos2\theta, \sin2\theta) .\qquad (7\text{-}14)$$

We note that the rays coming from a circle of radius ρ in the exit pupil lie on a circle of radius $2FA_c\rho^2$ in the image plane which is centered at $(4FA_c\rho^2, 0)$. The circle in the image plane is traced out twice as θ varies from 0 to 2π to complete a circle of rays in the exit pupil. Figure 7-2 illustrates these circles in the image plane for $\rho = 1/2$ and 1. For $\rho = 1/2$, the rays in the image plane lie on a circle of radius $FA_c/2$ centered at $(FA_c, 0)$. Accordingly, $AB/AP' = 1/2$, where P' is the Gaussian image point, so that the angle $AP'B$ is equal to 30°. Hence, all of the rays in the image plane are contained in a cone of semiangle 30° bounded by a circle of radius $2FA_c$ centered at $(4FA_c,0)$ corresponding to the marginal rays. The vertex of the cone, of course, coincides with the Gaussian image point P'. Since the spot diagram has the shape of a comet, the aberration is appropriately called coma. Note that the two tangential marginal rays ($\rho = 1$, $\theta = 0,\pi$) intersect this plane at T at a distance $6FA_c$ from P', and the two sagittal marginal rays ($\rho = 1$, $\theta = \pi/2, 3\pi/2$) intersect the image plane at S at a distance $2FA_c$ from P'. Accordingly, the length $6FA_c$ and half-width $2FA_c$ of the coma pattern are called *tangential* and *sagittal coma*, respectively.

7.5 Astigmatism

The astigmatism wave aberration is given by

$$W(\rho,\theta) = A_a\rho^2\cos^2\theta = A_a\xi^2 . \tag{7-15}$$

The corresponding ray aberrations are given by

$$(x_i,y_i) = (4FA_a\,\rho\cos\theta,0) = (4FA_a\xi,0) . \tag{7-16}$$

The point of intersection of a ray with the Gaussian image plane depends only on its ξ coordinate in the exit pupil. Thus, as indicated in Figure 7-3, all the rays transmitted by the exit pupil intersect the Gaussian image plane on a line along the x axis centered on the Gaussian image point. Since $-1 \le \xi \le 1$, the full width of this line is $8FA_a$.

If we add a small amount of defocus A_d to the astigmatism given by Eq. (7-15) by observing the image in a slightly defocused plane, the wave aberration becomes

$$W(\rho,\theta) = A_a\rho^2\cos^2\theta + A_d\rho^2 = (A_a + A_d)\xi^2 + A_d\eta^2 . \tag{7-17}$$

The corresponding ray aberrations are given by

$$(x_i,y_i) = 4F\,[(A_a + A_d)\rho\cos\theta, A_d\,\rho\sin\theta\,] \tag{7-18a}$$

$$= 4F\,[(A_a + A_d)\xi, A_d\eta] . \tag{7-18b}$$

For a given value of ρ, the locus of the points of intersection of the rays in the defocused image plane is given by

$$\left(\frac{x_i}{A}\right)^2 + \left(\frac{y_i}{B}\right)^2 = 1 , \tag{7-19}$$

where

$$A = 4F(A_a + A_d)\rho \text{ and } B = 4FA_d\rho . \tag{7-20}$$

Thus, the rays lying on a circle of radius ρ in the exit pupil, in general, lie in a defocused image plane on an ellipse whose semiaxes are given by A and B, respectively. The largest ellipse is obtained for the marginal rays. The relationship

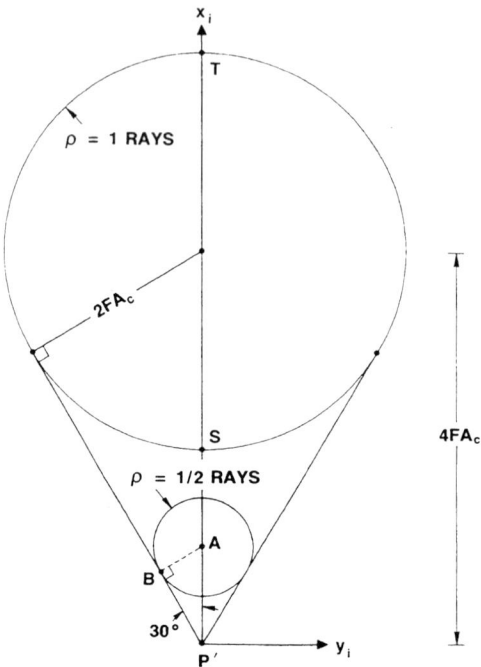

**Figure 7–2. Ray spot diagram for coma. All the rays lie in a cone of semiangle 30°
with vertex at the Gaussian image point P' bounded by a circle of radius $2FA_c$ centered
at $(4FA_c,0)$.**

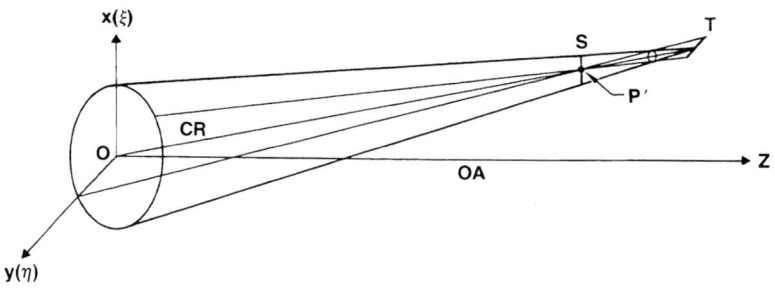

**Figure 7–3. Astigmatic focal lines. S is the sagittal focal line and T is the tangential
focal line. The rays in the tangential plane xz are focused at a point on T. Similarly,
the rays in the sagittal plane yz are focused at a point on S. The focal lines S and T
lie in the tangential and sagittal planes, respectively. For high quality imaging systems,
the field of view, and therefore the angle between OP' and the optical axis OA, is small.**

defocus wave aberration A_d and the corresponding longitudinal defocus is given by Eq. (7-11).

We note that if $A_d = 0$, the ellipse reduces to a line of full width of $8FA_a$ along the x axis. Thus, as discussed above, the image in the Gaussian image plane is a line S along the x axis centered on the Gaussian image point. If, however, $A_d = -A_a$, corresponding to $\Delta R = -8F^2A_a$, then the ellipse reduces to a line T along the y axis. The full width of this line image is the same as that of the line image S. The line image along the x axis is called the *sagittal* (or *radial*) *image* and lies in the tangential (or meridional) plane xz, containing the point object (which lies along the x axis in the object plane) and the optical axis. Similarly, the line image T along the y axis is called the *tangential image* and lies in the sagittal plane yz. The distance $8F^2A_a$ between the two line images is called *longitudinal astigmatism*. The two line images are called the *astigmatic focal lines*.

If $A_d = -A_a/2$, corresponding to $\Delta R = -4F^2A_a$, the ellipse reduces to a circle of maximum diameter of $4FA_a$, which is half the full width of the two line images. Since this circle is the smallest of all the possible images, Gaussian or defocused, it is called the *circle of least (astigmatic) confusion*.

Since $A_a \sim h'^2$, the width of the line images of a point object increases quadratically with the height h' of the Gaussian image point. Similarly, longitudinal astigmatism $8F^2A_a$ increases as h'^2. Thus, if we consider a line object, its sagittal image will also be a line which is slightly longer than (by an amount $8FA_a$) but coincident with its Gaussian image. However, its tangential image will be parabolic with a vertex radius of curvature of $h'^2/16F^2A_a$ or $1/4R^2a_a$. Similarly, the sagittal image of a planar object will be planar, but its tangential image will be paraboloidal. Note that longitudinal astigmatism corresponding to a Gaussian image at a height h' represents the sag of the tangential image surface at that height.

7.6   Field Curvature

The wave aberration corresponding to field curvature is given by

$$W(\rho) = A_d\rho^2 = A_d(\xi^2 + \eta^2) . \qquad (7-21)$$

Since the wave aberration is radially symmetric, the distribution of rays in the Gaussian image plane is also radially symmetric. For rays lying on a circle of radius ρ in the exit pupil, the radius of the circle of corresponding rays in the image plane, following Eq. (7-6), is given by

$$r_i = 4FA_d\rho . \qquad (7-22)$$

Its maximum value is $4FA_d$ and corresponds to the marginal rays.

From the discussion in Section 1.4, we note that a defocus aberration represented by Eq. (7-21) implies that the wavefront is spherical, but it is not centered at the Gaussian image point. Instead, it is centered at a distance

$$\Delta R = 8F^2 A_d \qquad (7\text{-}23)$$

from the Gaussian image point along the optical axis. (Strictly speaking, it is centered on the line joining the center of the exit pupil and the Gaussian image point.) Since the aberration coefficient $A_a \sim h'^2$, the sagittal image of a line object will be parabolic with a vertex radius of curvature of $h'^2/16F^2 A_d$, or $1/4R^2 a_d$. Similarly, the image of a planar object will be paraboloidal. The paraboloidal surface for a system with zero astigmatism is called the *Petzval image surface*.

7.7 Astigmatism and Field Curvature

Now we consider the combined effect of astigmatism and field curvature. Thus, the aberration with respect to the Gaussian image point is now given by

$$W(\rho,\theta) = A_a \rho^2 \cos^2\theta + A_d \rho^2 = (A_a + A_d)\xi^2 + A_d \eta^2 . \qquad (7\text{-}24)$$

Note that whereas in Eq. (7-17) the defocus coefficient was a variable, here it is fixed for a given point object. Since both A_a and A_d are proportional to h'^2, we find, following the discussion of Sections 7.5 and 7.6, that the sagittal and tangential images of a line object are formed on parabolic curves with vertex radii of curvature given by

$$R_s = h'^2/16F^2 A_d = 1/4R^2 a_d \qquad (7\text{-}25)$$

and

$$R_t = h'^2/16F^2(A_a + A_d) = 1/4R^2(a_a + a_d) , \qquad (7\text{-}26)$$

respectively. Similarily, the images of a planar object centered on the optical axis will be the corresponding paraboloids symmetric about the optical axis.

Combining Eqs. (7-25) and (7-26) with Eq. (1-28), for imaging by a spherical refracting surface, where L is the same as R here, we find that

$$\frac{3}{R_s} - \frac{1}{R_t} = \frac{2}{R_p} . \qquad (7\text{-}27)$$

It has the consequence that the Petzval surface is three times as far from the tangential surface as it is from the sagittal surface, as may be seen by comparing the sags of the three surfaces. Moreover, the sagittal surface always lies between the tangential and the Petzval surfaces. When astigmatism is zero, the sagittal and tangential surfaces coincide with the Petzval surface. Although Eq. (7-27) and its consequences have been obtained for a single spherical refracting surface, they hold for any rotationally symmetric imaging system.

7.8   Distortion

The distortion wave aberration is given by

$$W(\rho,\theta) = A_t \rho\cos\theta = A_t \xi ,\qquad (7\text{-}28)$$

where the aberration coefficient A_t is proportional to h'^3. The corresponding ray aberrations are given by

$$(x_i,y_i) = (2FA_t,0) .\qquad (7\text{-}29)$$

Since the ray aberrations are independent of the coordinates (ρ,θ) of a ray in the exit pupil, all the rays converge at the image point $(2FA_t,0)$, which lies along the x axis at a distance $2FA_t$ from the Gaussian image point. Thus, a wavefront aberrated by distortion is tilted with respect to the Gaussian reference sphere by an angle

$$\beta = A_t/a .\qquad (7\text{-}30)$$

This angle is proportional to h'^3. Similarly, the distance $2FA_t$ of the perfect image point from the Gaussian image point is proportional to h'^3. This gives rise to the familiar *pincushion* or *barrel* distorted image of a square grid, depending on whether A_t is positive or negative, respectively.

PART II

Wave Diffraction Optics

CHAPTER 8

Optical Systems With Circular Pupils

8.1 Introduction

In this chapter, we consider optical systems with *circular exit pupils* and discuss imaging by them based on the diffraction of object radiation at the exit pupil. Our starting point is an equation for the distribution of light in the image of a point object called the *diffraction point-spread function* (PSF) of the system. This equation is equally suitable for calculating the *diffraction pattern* of a circular aperture. Since, under certain conditions, the *diffraction image* of an *incoherent object* is given by the convolution of its Gaussian image (which is a scaled replica of the object) and the system PSF,[1,2] the PSF calculations are fundamental to the theory of optical imaging. To understand the effect of aberrations on images, it is essential that we first understand the aberration-free PSF. Accordingly, we give briefly the characteristics of the aberration-free image of a point object. Our discussion on aberrated images is built slowly. First, we discuss *defocused images* and irradiance along the axis of the pupil. Next, approximate relationships between the ratio of the PSF values at its center with and without aberration, called the *Strehl ratio*, and the variance of the aberration across the pupil are developed. The approximate results for *primary aberrations* are compared with the corresponding exact results to determine the range of validity of the simple Strehl ratio formulas. The concept of *aberration balancing* is introduced in which an aberration of a certain order in pupil coordinates is mixed or balanced with one or more aberrations of lower order to minimize its variance, and thereby maximize the Strehl ratio of the system. *Aberration tolerances* based on a Strehl ratio of 0.8 are given for primary and *balanced primary aberrations*. Rayleigh's *quarter-wave rule* is briefly discussed and balanced primary aberrations are identified with *Zernike circle polynomials*. Finally, aberrated PSFs for various amounts of primary aberrations are given and their *symmetry properties* in and about the Gaussian image plane are discussed.

Since the diffraction image of an incoherent isoplanatic object is given by the convolution of its Gaussian image and the PSF of the system forming the image, the *spatial frequency spectrum* of the diffraction image is given by the product of the spectrum of the Gaussian image and the *optical transfer function* (OTF) of the system.[1] Thus, the OTF of the system, which is equal to the Fourier transform of its PSF, is equally fundamental to the theory of optical imaging. The OTF of an aberration-free system with a circular pupil is given, and how it is affected by an aberration is discussed. The concept of a *Hopkins ratio*, representing the ratio of the magnitudes of the OTFs at a certain spatial frequency, with and without aberration is introduced. It is related to the variance of an *aberration difference function* across the region of overlap between two displaced pupils, the displacement depending on the spatial frequency under consideration. Aberration tolerances based on a Hopkins ratio of 0.8 are given for primary aberrations. Finally, the OTFs for various amounts of primary aberrations are discussed.

Only the monochromatic PSF and OTF are discussed here, although we point out how these functions depend on the optical wavelength. How one may calculate these functions for "white" light is also briefly mentioned.

8.2. Point-Spread Function (PSF)

In this section we give a general equation for evaluating the aberrated PSF of a system with a circular exit pupil. We give closed-form analytical solutions for the aberration-free PSF and encircled power giving the fraction of total power in the image of a point object in a circle of some radius centered at the Gaussian image point. Defocused PSFs and axial irradiance of the convergent image-forming beam are considered next. It is shown, for example, that the irradiance distribution even for an aberration-free system is not symmetric about the Gaussian image plane unless the Fresnel number of the pupil (defined below) as observed from the Gaussian image point is very large. The content of this section forms the basis from which to study the effects of aberrations on the images.

8.2.1 Aberrated PSF

Consider an aberrated optical system with a circular exit pupil of radius a imaging a point object radiating at a wavelength λ. Let R be the distance between the planes of the exit pupil and the Gaussian image and let $\Phi(\rho,\theta)$ be the *phase aberration* at a point (ρ,θ) in the pupil plane, where ρ is in units of a. The phase aberration Φ is related to the *wave aberration* $W(\rho,\theta)$ considered in earlier chapters according to $\Phi = (2\pi/\lambda)W$. The diffraction PSF or the irradiance distribution of the image in a plane normal to its optical or the z axis at a distance z from the plane of its exit pupil may be written[1]

$$I(r, \theta_i; z) = \frac{PS_p}{\pi^2 \lambda^2 z^2} \left| \int_0^1 \int_0^{2\pi} \exp[i\Phi(\rho,\theta)] \exp[-\pi i \frac{R}{z} r \rho \cos(\theta - \theta_i)] \rho \, d\rho \, d\theta \right|^2, \quad (8\text{-}1)$$

where (r,θ_i) are the polar coordinates of the observation point with respect to the point where the line joining the center of the exit pupil and the Gaussian image point intersects the observation plane, r is in units of $\lambda R/2a = \lambda F$ (F being the focal ratio or the f-number of the image-forming light cone), P is the total power in the exit pupil and therefore in the image, and $S_p = \pi a^2$ is the area of the exit pupil of the system. Strictly speaking, the PSF of a system represents the irradiance distribution of the image of a point object per total power in the image. Accordingly, whereas the irradiance is in units of W/m², the PSF is in units of m⁻². The angles θ and θ_i are zero for pupil and observation points lying in the tangential plane on the positive side of the x axis. As in earlier chapters, we assume that the point object lies along the x axis so that the zx plane represents the tangential plane.

The function $\exp[i\Phi(\rho,\theta)]$ is called the *pupil function* of the system. A system whose aberration function $\Phi(\rho,\theta)$ is (approximately) the same for all points on an extended object is called *isoplanatic*. The image of an incoherent object formed by such a system is obtained by convolving its Gaussian image with the PSF of the system, i.e., by adding the irradiance distributions of its image elements. Similarly, the complex amplitude distribution of the image of a *coherent object* formed by such a system is obtained by adding the complex amplitude distributions of its image elements.

The PSF of a system depends on the optical wavelength λ in several ways. Referring to Eq. (8-1), first, the power P in the exit pupil at a certain wavelength (strictly speaking, across a narrow spectral band with a certain mean wavelength) may be different from that at another wavelength. This variation will depend on the spectral radiance distribution of the object and the spectral transmission of the system. Second, there is an inverse-square-law dependence on the wavelength. It impacts the "brightness" of the PSF; the shorter the wavelength, the brighter the PSF. Third, the wave aberration may depend on the wavelength if the system has one or more dispersive elements. Even if the system is nondispersive, the phase aberration is inversely proportional to it. Hence, the effect of an aberration on the PSF is different at two different wavelengths. Fourth, the variable r is normalized by the wavelength. It impacts the "size" of the PSF; the shorter the wavelength, the narrower the PSF. However, a shorter wavelength also means larger phase aberration and, therefore, more spreading of the PSF due to the aberration. The white light or polychromatic PSF may be determined by integrating the monochromatic PSF across the spectral distribution of the image forming radiation.

8.2.2 Aberration-Free PSF

It can be shown that the aberration-free irradiance distribution, obtained from Eq. (8-1) by letting $\Phi(\rho,\theta) = 0$ and $z = R$, is given by

$$I(r;R) = \frac{PS_p}{\lambda^2 R^2} \left[\frac{2J_1(\pi r)}{\pi r} \right]^2 \, , \tag{8-2}$$

where $J_1(\bullet)$ is the first-order Bessel function of the first kind. This distribution normalized by its central value $PS_p/\lambda^2 R^2$ and shown in Figure 8-1 is called the *Airy pattern*.[2] The power contained in a circle of radius r_c (in units of λF) centered at the Gaussian image point $r = 0$ is given by

$$P(r_c;R) = P[1 - J_0^2(\pi r_c) - J_1^2(\pi r_c)] \, . \tag{8-3}$$

The encircled-power distribution normalized by the total power P is also shown in Figure 8-1. The location of the *maxima* and *minima* of the irradiance distribution, the values of irradiance at these points, and the corresponding encircled powers are given in Table 8-1. The minima and maxima correspond to the roots of $J_1(\pi r) = 0$ and $J_2(\pi r) = 0$, respectively, where $J_2(\bullet)$ is the second-order Bessel function of the first kind. It should be evident that the encircled power corresponding to minima is given by $1 - J_0^2(\pi r_m)$, where r_m represents the value of r for a minimum. The central bright spot of radius 1.22 containing 83.8% of the total power is called the *Airy disc*. Note that the

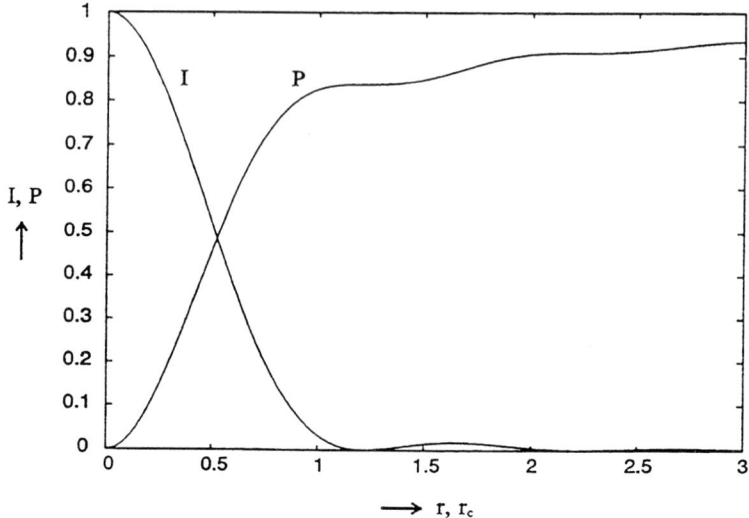

Figure 8–1. Irradiance and encircled-power distributions of an Airy pattern.

principal maximum of the irradiance distribution lies at $r = 0$ where Huygens' spherical wavelets originating at the exit pupil interfere constructively. The aberration-free image of an object is also called its *diffraction-limited image* (since the quality of the image is limited only by diffraction of the object radiation at the exit pupil of the system). It should be noted that the radius of the Airy disc increases linearly with wavelength and the irradiance at its center decreases quadratically with it.

8.2.3 Rotationally Symmetric PSF

For a radially symmetric aberration $\Phi(\rho)$, Eq. (8-1) reduces to

$$I(r;z) = \frac{4PS_p}{\lambda^2 z^2} \left| \int_0^1 \exp[i\Phi(\rho)] \, J_0(\pi r \rho R/z)\rho d\rho \right|^2 . \tag{8-4}$$

It is clear from Eq. (8-4) that the irradiance distribution is rotationally symmetric about the z axis. Hence, it is radially symmetric in any observation plane normal to it. Moreover, it does not depend on the sign of $\Phi(\rho)$ (as may be seen by changing i to $-i$).

8.2.4 Defocused PSF

If the imaging system is aberration free but the image is observed in a plane $z \neq R$, then the image suffers from defocus aberration given by

$$\Phi(\rho) = A_d \rho^2 , \tag{8-5}$$

where

$$A_d = \frac{\pi a^2}{\lambda} \left(\frac{1}{z} - \frac{1}{R} \right) \tag{8-6a}$$

$$= \pi N \left(\frac{R}{z} - 1 \right) , \tag{8-6b}$$

Table 8–1. Irradiance and encircled power corresponding to the maxima and minima of the irradiance distribution of the image of a point object formed by an optical system with a circular exit pupil. The irradiance is in units of the irradiance $I(0)$ at the Gaussian image point and the encircled power is in units of the total power P in the exit pupil and, therefore, in the image. r and r_c are in units of λF.

Max/Min	r, r_c	$I(r)$	$P(r_c)$
Max	0	1	0
Min	1.22	0	0.838
Max	1.64	0.0175	0.867
Min	2.23	0	0.910
Max	2.68	0.0042	0.922
Min	3.24	0	0.938
Max	3.70	0.0016	0.944
Min	4.24	0	0.952
Max	4.71	0.0008	0.957

is the peak defocus aberration. In Eq. (8-6b), $N = a^2/\lambda R$ is the *Fresnel number* of the exit pupil as observed from the center of the Gaussian image plane. Thus, the edge of the exit pupil is farther than its center by approximately $N\lambda/2$ from the center of the Gaussian image plane so that N is the number of Fresnel's *half–wave zones* in the exit pupil.

From Eq. (8-1) we note that the irradiance distribution is asymmetric about the plane $z = R$; i.e., the distributions in two planes located at $z = R \pm \Delta$, where Δ is a longitudinal defocus, are not identical, even if the system is otherwise aberration free. There are three reasons for this asymmetry. First, the inverse-square law dependence on z increases the irradiance for $z < R$ and decreases it for $z > R$. Second, A_d is asymmetric since the defocus coefficients for these two planes are different, as may be seen from Eq. (8-6).Third, the exponent in Eq. (8-1), which determines the scale of the image, depends on z.

For systems with a small Fresnel number ($N \leq 10$), z can be much different from R for A_d to achieve a significant value. Accordingly, all of the three factors mentioned above contribute to the asymmetry of the irradiance distribution about the plane $z = R$. One consequence of this is that the irradiance at points on and near the z axis can be higher for $z < R$ than for $z = R$. For example, a beam of diameter 25 cm and a wavelength of 10.6 μm focused at a distance of 1.47 km corresponds to $N = 1$. A Strehl ratio [discussed in the next section and whose exact value is given by the square of the quantity in parenthesis in Equation (8-7)] of 0.8 is obtained at two z values: 0.97 km and 3.03 km. The principal maximum of axial irradiance occurs at a distance of $0.6R = 0.88$ km.

If the Fresnel number of a system is very large ($N >> 10$), A_d becomes large even for very small differences in z and R. For example, a photographic system with $a =$

1 cm, $\lambda = 0.5\,\mu\text{m}$, and $R = 10$ cm corresponds to $N = 2000$, and a Strehl ratio of 0.8 is obtained for $z = R \pm 25\,\mu\text{m}$. Accordingly, the defocus tolerance for such systems dictates that z be practically equal to R. Hence, following Eq. (8-6a), we note that two observation planes $z = R \pm \Delta$ correspond to defocus coefficients of $A_d = \mp\,\pi\Delta/4\lambda F^2$. Since these coefficients are equal in magnitude but opposite in sign, letting $\Phi(\rho) = A_d\rho^2$ in Eq. (8-4), we find that the irradiance distribution for an unaberrated system with a large Fresnel number is symmetric about the Gaussian image plane $z = R$.

8.2.5 Axial Irradiance

If we let $r = 0$ in Eq. (8-1), we obtain the irradiance along the z axis (strictly speaking, along the line joining the center of the exit pupil and the Gaussian image point). For an aberration–free system, the axial irradiance is given by

$$I(0; z) = \frac{PS_p}{\lambda^2 z^2}\left(\frac{\sin A_d/2}{A_d/2}\right)^2 . \qquad (8\text{-}7)$$

Figure 8-2 shows how this irradiance varies with z for systems with $N = 1$, 10, and 100.[3] We note that it is highly asymmetric about the Gaussian image point ($z = R$) when $N = 1$, but it becomes more and more symmetric as N increases. The axial irradiance for a Gaussian pupil (with a truncation parameter of unity discussed in Chapter 9) is also shown in this figure with a similar behavior. It should be noted that even though the principal maximum of axial irradiance does not lie at the focus, unless N is very large, maximum central irradiance on a target in a beam-focusing system always occurs when the beam is focused on it.[3]

8.3   Strehl Ratio

Now we consider an aberrated system and discuss how the value at the center of the PSF is affected by the aberration in the system, thereby introducing the concept of the Strehl ratio. We obtain simple but approximate expressions for the Strehl ratio in terms of the variance of the aberration across the exit pupil of the system. We introduce the concept of aberration balancing in which an aberration of a certain order in pupil coordinates is balanced by one or more aberrations of lower order to minimize its variance and thus maximize the Strehl ratio. We give aberration tolerances for primary and balanced primary aberrations corresponding to a Strehl ratio of 0.8. A brief discussion of Rayleigh's quarter-wave rule is given, and balanced primary aberrations are identified with some of the Zernike circle polynomials.

8.3.1 General Expressions

The *Strehl irradiance ratio* of an image or a system, defined as the ratio of the irradiance values at the center of the image in a plane with and without the aberration, according to Eq. (8-1) is given by

$$S = \pi^{-2}\left|\int_0^1\int_0^{2\pi}\exp[i\Phi(\rho,\theta)]\rho\,d\rho\,d\theta\right|^2 \qquad (8\text{-}8)$$

$$= \left|<\exp\left[i(\Phi - <\Phi>)\right]>\right|^2$$

$$= <\cos\,(\Phi - <\Phi>)>^2 + <\sin\,(\Phi - <\Phi>)>^2$$

$$\geq <\cos(\Phi - <\Phi>)>^2 , \qquad (8\text{-}9)$$

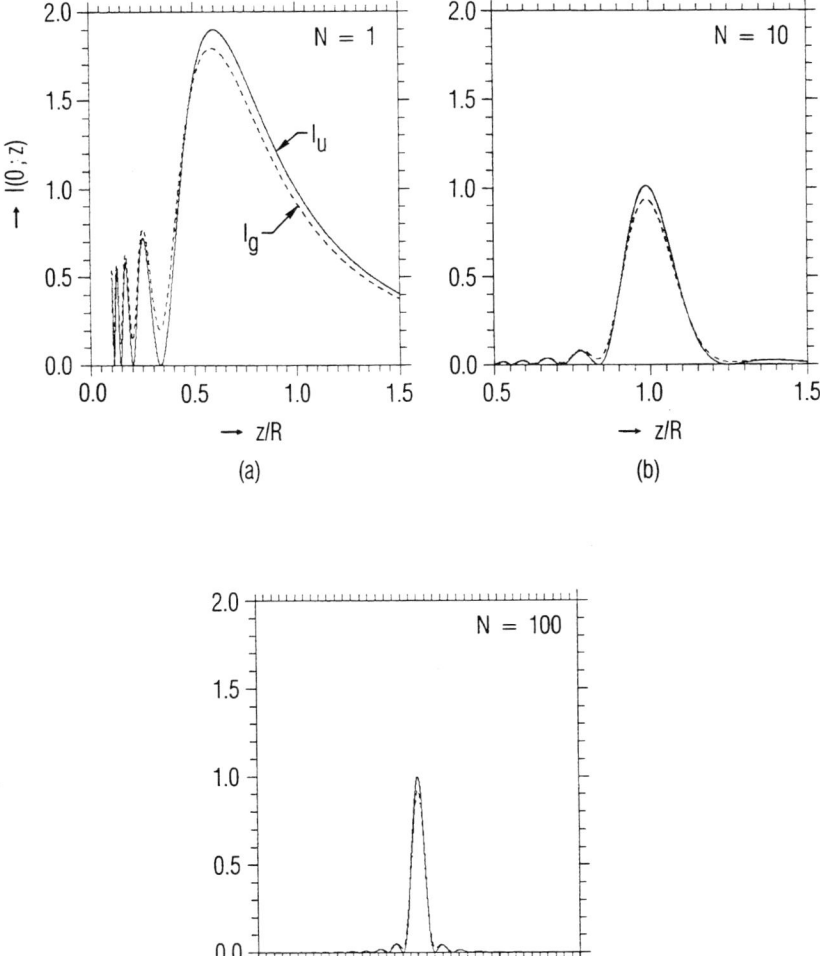

Figure 8-2. Axial irradiance of a circular beam focused at a distance R with a Fresnel number $N = a^2/\lambda R = 1, 10, 100$. The irradiance is in units of the focal-point irradiance $PS_p/\lambda^2 R^2$. The subscripts u and g refer to uniform and (truncated) Gaussian beams, respectively. Gaussian beams are discussed in Chapter 9.

where the angular brackets indicate an average across the pupil. Expanding the cosine function in a power series and retaining the first two terms for small aberrations yields the Maréchal result[4]

$$S \geq (1 - \sigma_\Phi^2/2)^2 , \qquad (8\text{-}10)$$

where

$$\sigma_\Phi^2 = <(\Phi - <\Phi>)^2>$$

$$= <\Phi^2> - <\Phi>^2 \qquad (8\text{-}11)$$

is the *variance* of the phase aberration across the pupil. Note that

$$<\Phi^n> = \pi^{-1} \int_0^1 \int_0^{2\pi} \Phi^n(\rho,\theta) \rho \, d\rho \, d\theta . \qquad (8\text{-}12)$$

For small values of *standard deviation* σ_Φ three approximate expressions have been used in the literature:

$$S_1 \simeq (1 - \sigma_\Phi^2/2)^2 \qquad (8\text{-}13)$$

$$S_2 \simeq 1 - \sigma_\Phi^2 , \qquad (8\text{-}14)$$

and

$$S_3 \simeq \exp(-\sigma_\Phi^2) . \qquad (8\text{-}15)$$

The first is the *Maréchal formula*, the second is the commonly used expression obtained when the term in σ_Φ^4 in the first is neglected,[5] and the third is an empirical expression giving a better fit to the actual numerical results for various aberrations.[6] We note that the Strehl ratio for a small aberration does not depend on its type but only on its variance across the pupil.

8.3.2 Primary Aberrations

Table 8-2 gives the form as well as the standard deviation σ_Φ of a primary aberration. Here, the aberration coefficient A_i represents the peak value of an aberration. (The balanced aberrations noted in the table are considered in the next section.) Comparing it with the aberration coefficients a_i considered in earlier chapters, we note, for example, that $A_c = a_c h' a^3$, where h' is the height of the Gaussian image point from the optical axis of the system. It also lists the tolerance for an aberration coefficient A_i for a Strehl ratio of 0.8. The *optical tolerances* listed in Table 8-2 are for the wave (as opposed to phase) aberration coefficient, as is customary in optics. A Strehl ratio of 0.8 corresponds to an aberration with a standard deviation of $\sigma_w = \lambda/14$.

8.3.3 Balanced Primary Aberrations

In Chapter 7, where we discussed ray aberrations, we mixed one aberration with another in order to minimize the size of the ray spot in an image plane. For example, in the case of spherical aberration, the circle of least confusion was determined to be in a plane 3/4 of the way from the paraxial (Gaussian) image plane to the marginal image plane. The radius of the circle of least confusion was found to be 1/4 of the spot radius in the paraxial plane. Similarly, in the case of astigmatism, the circle of least confusion was determined to be in a plane lying midway between the planes containing the sagittal (Gaussian) and tangential line images. This circle had a diameter which was half the length of the line images.

Table 8–2. Primary aberrations, their standard deviations, and values of aberration coefficients, peak aberrations, and peak-to-valley aberrations for a Strehl ratio of 0.8.

Aberration	$\Phi(\rho,\theta)$	σ_Φ	W_p	$W_{p\text{-}v}$	$S = 0.8$		
					A_i	W_p	$W_{p\text{-}v}$
Spherical	$A_s\rho^4$	$\dfrac{2A_s}{3\sqrt{5}}$	A_s	A_s	0.25	0.25	0.25
Balanced Spherical	$A_s(\rho^4 - \rho^2)$	$\dfrac{A_s}{6\sqrt{5}}$	$A_s/4$	$A_s/4$	1	0.25	0.25
Coma	$A_c\rho^3\cos\theta$	$\dfrac{A_c}{2\sqrt{2}}$	A_c	$2A_c$	0.21	0.21	0.42
Balanced Coma	$A_c(\rho^3 - 2\rho/3)\cos\theta$	$\dfrac{A_c}{6\sqrt{2}}$	$A_c/3$	$2A_c/3$	0.63	0.21	0.42
Astigmatism	$A_a\rho^2\cos^2\theta$	$\dfrac{A_a}{4}$	A_a	A_a	0.30	0.30	0.30
Balanced Astigmatism	$A_a\rho^2(\cos^2\theta - 1/2)$ $= (A_a/2)\rho^2\cos2\theta$	$\dfrac{A_a}{2\sqrt{6}}$	$A_a/2$	A_a	0.37	0.18	0.37

For small aberrations, since the Strehl ratio is maximum when the aberration variance is minimum, the best image plane is one which corresponds to minimum variance. Thus, for example, we balance spherical aberration with defocus and write it as

$$\Phi(\rho) = A_s\rho^4 + A_d\rho^2 . \tag{8-16}$$

We determine the amount of defocus A_d such that the variance σ_Φ^2 is minimized; i.e., we calculate σ_Φ^2 and let

$$\frac{\partial\sigma_\Phi^2}{\partial A_d} = 0 \tag{8-17}$$

to determine A_d. Proceeding in this manner, we find that the optimum value is $A_d = -A_s$. Figure 8-3 shows how the aberration varies with ρ for this value of A_d and for $A_d = 0$. The standard deviation of the optimally balanced aberration is $A_s/6\sqrt{5}$, which is a factor of 4 smaller than the standard deviation for $A_d = 0$. Since the standard deviation has been reduced by a factor of 4 by balancing spherical aberration with defocus, the optical tolerance has been increased by the same factor. Following Section 1.4, a defocus of $A_d = -A_s$ is introduced by observing the image in a plane which is farther from the exit pupil than the Gaussian image plane by $8F^2A_s$. Moreover, since $A_d = 0$ and $A_d = -2 A_s$ correspond to paraxial and marginal image planes, respectively, we note that, based on diffraction, the best image is obtained in a plane lying midway between them. This is different from the plane containing the circle of least confusion which corresponds to $A_d = -1.5 A_s$.

Coma and astigmatism can be treated similarly. Table 8-2 lists the form of a balanced primary aberration, its standard deviation, and its tolerance for a Strehl ratio of 0.8. We note that in the case of coma, the balancing aberration is a wavefront tilt with a coefficient that is minus two-thirds of the coma coefficient. Thus, maximum Strehl ratio is obtained at a point which is displaced from the Gaussian image point by $(4FA_c/3,0)$ but lies in the Gaussian image plane. By balancing coma with an appropriate amount of tilt, its standard deviation is reduced by a factor of 3. In the case of astigmatism, the best Strehl ratio is obtained in a plane which is farther than the Gaussian image plane by $4F^2A_a$. As discussed in Chapter 7, this is also the plane in which the circle of least confusion is obtained. By balancing with defocus, the standard deviation of astigmatism is reduced by a factor of 1.225. The point of observation with respect to which the aberration variance is minimum and, therefore, the irradiance at that point is maximum, is called the *diffraction focus*.

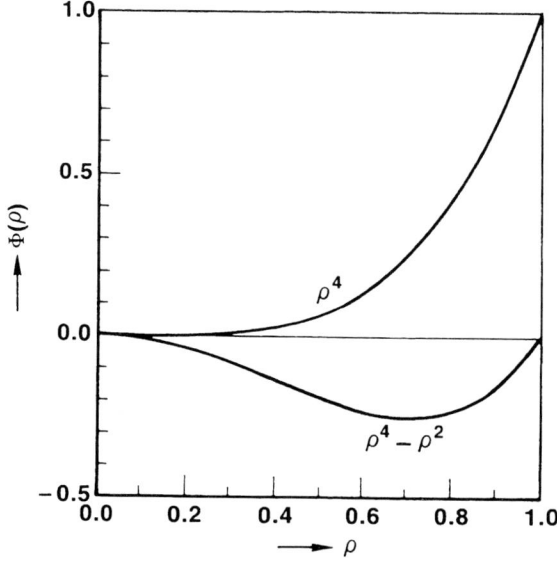

Figure 8–3. Variation of spherical and balanced spherical aberrations with radial variable ρ.

8.3.4 Comparison of Approximate and Exact Results

Figure 8-4 shows how the Strehl ratio of a primary aberration varies with its standard deviation. Approximate as well as exact results are shown in this figure.[6] The exact results are obtained by the use of Eq. 8-8. The curves for a given aberration and for the corresponding balanced aberration can be distinguished from each other by their behavior for large σ_w values (near 0.25λ). For example, coma is shown by the evenly dashed curves; the higher dashed curve is for coma and the lower is for balanced coma. The same holds true for astigmatism. The curves for spherical and balanced spherical aberrations are identical since the Strehl ratio for a given value of σ_w is the same for the two aberrations. The following observations may be made from Figure 8-4:

i.   For small values of σ_w, the Strehl ratio is independent of the type of aberration. It depends only on its variance.

ii.   The expressions for S_1 and S_2 underestimate the true Strehl ratio S.

iii.   The expression for S_3 underestimates the true Strehl ratio only for coma and astigmatism; it overestimates for the other aberrations. Numerical analysis shows that the error, defined as $100(1 - S_3/S)$, is $< 10\%$ for $S > 0.3$.

iv. S_3 gives a better approximation for the true Strehl ratio than S_1 and S_2. The reason is that, for small values of σ_Φ, it is larger than S_1 by approximately σ_Φ^4. Of course, S_1 is larger than S_2 by $\sigma_\Phi^4/4$.

v.   The Strehl ratio depends strongly on the standard deviation of an aberration but weakly on its detailed distribution over a wide range of Strehl ratio values.

8.3.5 Strehl Ratio for Nonoptimally Balanced Aberrations

When a certain aberration is balanced with other aberrations to minimize its variance, the balanced aberration does not necessarily yield a higher or the highest possible Strehl ratio. For small aberrations, a maximum Strehl ratio should be obtained according to any of the Eqs. (8-13)–(8-15), when the variance is minimum. For large aberrations, however, there is no simple relationship between the Strehl ratio and the aberration variance. For example,[7] when $A_s = 3\lambda$, the optimum amount of defocus is $A_d = -3\lambda$, but the Strehl ratio is a minimum and equal to 0.12. The Strehl ratio is maximum and equal to 0.26 for $A_d \approx 4\lambda$ or -2λ. For $A_s \lesssim 2.3\lambda$, the axial irradiance is maximum at a point with respect to which the aberration variance is minimum. Similarly, in the case of coma, the maximum irradiance in the image plane occurs at the point with respect to which the aberration variance is minimum only if $A_c \lesssim 0.7\lambda$, which in turn corresponds to $S \gtrsim 0.76$. For larger values of A_c, the distance of the point of maximum irradiance does not increase linearly with its value and even fluctuates in some regions.[8] Moreover, it is found that for $A_c > 2.3\lambda$, the Seidel coma gives a larger Strehl ratio than the balanced coma; i.e., the irradiance in the image plane at the origin is larger than at the point with respect to which the aberration variance is minimum. Thus, only for large Strehl ratios, the irradiance is maximum at the point associated with minimum aberration variance.

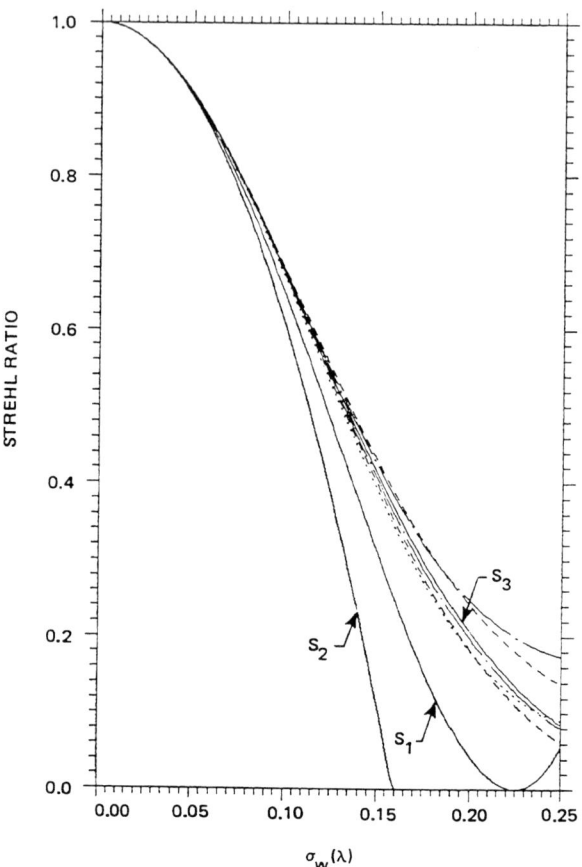

Figure 8–4. Strehl ratio for primary aberrations as a function of their standard deviation σ_w in units of optical wavelength λ. For large values of σ_w, coma and astigmatism give a higher Strehl ratio than the corresponding balanced aberration. The Strehl ratios for spherical and balanced spherical aberrations are identical. $\sigma_\Phi = (2\pi/\lambda)\sigma_w$. Spherical...., Coma——, Astigmatism –·– (from Reference 7).

When *secondary spherical aberration* (varying as ρ^6) and *secondary coma* (varying as $\rho^5\cos\theta$) are balanced with lower-order aberrations to minimize their variance, it is found[9] that a maximum of Strehl ratio is obtained only if its value comes out to be greater than about 0.5. Otherwise, a mixture of aberrations yielding a larger-than-minimum possible variance gives a higher Strehl ratio than the one provided by a minimum variance mixture.

8.3.6 Rayleigh's $\lambda/4$ Rule

Rayleigh[10] showed that a quarter-wave of primary spherical aberration reduces the irradiance at the Gaussian image point by 20%; i.e., the Strehl ratio for this aberration is 0.8. This result has brought forth *Rayleigh's* $\lambda/4$ *rule*, namely, that a Strehl ratio of approximately 0.8 is obtained if the maximum absolute value of the aberration at any point in the pupil is equal to $\lambda/4$. A variant of this definition is that an aberrated wavefront which lies between two concentric spheres that are spaced a quarter-wave apart will give a Strehl ratio of approximately 0.8. Thus, instead of $W_p = \lambda/4$, we require $W_{p-v} = \lambda/4$, where W_p is the peak absolute value and W_{p-v} is the peak-to-valley value of the aberration. From Table 8-2, we note that a Strehl ratio of 0.8 is obtained for $W_p = \lambda/4 = W_{p-v}$ for spherical aberration only. For other primary aberrations, distinctly different values of W_p and W_{p-v} give a Strehl ratio of 0.8. In Table 8-2, W_p and W_{p-v} are also given in terms of the aberration coefficient A_i. Thus, we see that it is advantageous to use σ_w for estimating the Strehl ratio. A Strehl ratio ≥ 0.8 is obtained for $\sigma_w \leq \lambda/14$.

8.3.7 Balanced Aberrations and Zernike Circle Polynomials

A *balanced aberration* represents an aberration in which an aberration of a certain order in the power series expansion of the aberration function in pupil coordinates is mixed with aberrations of lower order such that the variance of the net aberration is minimized. Consider, for example, a typical balanced aberration

$$\Phi_n^m(\rho, \theta) = c_{nm} \epsilon_m \sqrt{2(n + 1)} R_n^m(\rho) \cos m\theta , \qquad (8\text{-}18)$$

which represents a term in the expansion of the aberration function in terms of a complete set of *Zernike circle polynomials*,[1] which are orthonormal over a unit circular pupil, where n and m are positive integers (including zero), $n - m \geq 0$ and even. The radial polynomial

$$R_n^m(\rho) = \sum_{s=0}^{(n-m)/2} \frac{(-1)^s (n-s)!}{s![(n+m)/2-s]![(n-m)/2-s]!} \rho^{n-2s} \qquad (8\text{-}19)$$

is a polynomial of degree n in ρ containing terms in ρ^n, ρ^{n-2}, ..., and ρ^m. The quantity

$$\epsilon_m = 1/\sqrt{2} , m = 0$$

$$= 1 \quad , m \neq 0 . \qquad (8\text{-}20)$$

Unless $n = m = 0$, the coefficient c_{nm} represents the standard deviation of the aberration across the pupil, i.e.,

$$c_{nm} = \sigma_\Phi . \qquad (8\text{-}21)$$

The orthonormality of the Zernike polynomials implies that

$$\int_0^1 \int_0^{2\pi} \Phi_n^m(\rho,\theta)\Phi_{n'}^{m'}(\rho,\theta)\rho \, d\rho \, d\theta \Big/ \int_0^1 \int_0^{2\pi} \rho \, d\rho \, d\theta = c_{nm}^2 \delta_{nn'} \delta_{mm'} , \qquad (8\text{-}22)$$

where δ_{ij} is a Kronecker delta.

The balanced primary aberrations considered in Table 8-2 can be easily identified with the corresponding Zernike polynomials. Since

$$R_4^0(\rho) = 6\rho^4 - 6\rho^2 + 1 \qquad (8\text{-}23)$$

and

$$R_3^1(\rho) = 3\rho^3 - 2\rho \tag{8-24}$$

$$R_2^2(\rho) = \rho^2 , \tag{8-25}$$

we note that Φ_4^0, Φ_3^1, and Φ_2^2 represent balanced aberrations with

$$c_{40} = A_s/6\sqrt{5} \tag{8-26}$$

$$c_{31} = A_c/6\sqrt{2} \tag{8-27}$$

and

$$c_{22} = A_a/2\sqrt{6} . \tag{8-28}$$

The aberration Φ_4^0 contains a constant (independent of ρ and θ) term. This term does not change the standard deviation of the balanced aberration or the Strehl ratio corresponding to it. Similarly, Zernike polynomials Φ_2^0, Φ_1^1, and Φ_0^0 represent wavefront defocus, tilt, and *piston* (i.e., uniform aberration), respectively. The corresponding radial polynomials are given by $R_2^0(\rho) = 2\rho^2 - 1$, $R_1^1(\rho) = \rho$, and $R_0^0(\rho) = 1$, respectively.

Thus, we see that Zernike polynomials can be identified with balanced aberrations; that, in fact, is their advantage. For obvious reasons, balanced aberrations represented by Zernike polynomials may be referred to as *Zernike* or *orthogonal aberrations*. Thus, for example, Φ_4^0, Φ_3^1, and Φ_2^2 represent *Zernike spherical aberration, Zernike coma*, and *Zernike astigmatism* respectively. Here we have discussed only the primary aberrations. In general, the aberration function of an optical system may consist of higher-order aberrations. Moreover, in a system without an axis of rotational symmetry, the aberration function will consist of terms not only in $\cos m\theta$ but in $\sin m\theta$ as well. Because of the orthogonality of Zernike polynomials, the variance of an aberration is given by the sum of the variance of each of its orthogonal components.

8.4   PSFs for Primary Aberrations

In the previous section, we have discussed how the central value of the aberration-free PSF is affected by an aberration in the system. Now, we consider the full PSFs for primary aberrations. In particular, we discuss their symmetry properties in and about the Gaussian image plane. Pictures of the aberrated PSFs for various values of a primary aberration are also given, since that is what one would see in practice when observing the image of a point object formed by an aberrated system.

We have already shown in Section 8.2 that the PSF of an aberration-free system in any plane normal to its optical (z) axis is radially symmetric about the center. We also know that, unless the Fresnel number of the system is small ($N \leq 10$), the irradiance distribution of the image of a point object is symmetric about the plane $z = R$. For systems with large Fresnel numbers ($N >> 10$), certain additional symmetry properties can be obtained from Eq. (8-1) or its extensions. They are stated below and summarized in Table 8-3. The PSFs for various aberrations are also given in this section.

Table 8-3. Symmetry of PSFs aberrated by primary aberrations.

Aberration	General Symmetry	Symmetry of Axial Irradiance	Symmetry in Defocused Images	Symmetry in Coefficient Sign
None	Rotational about z axis Radial in any z plane	About $A_d = 0$	About $A_d = 0$	Not applicable
Spherical $A_s\rho^4$	Rotational about z axis Radial in any z plane	About $A_d = -A_s$	About $A_d = -A_s$ if $A_s \rightarrow -A_s$	In $A_d = 0$ plane
Astigmatism $A_a\rho^2\cos^2\theta$	4-fold in $A_d = -A_d/2$ plane, line about x and y axes	About $A_d = -A_a/2$	About $A_d = -A_a/2$ if rotated by $\pi/2$ or if $A_a \rightarrow -A_a$	In $A_d = -A_a/2$ plane
Coma $A_c\rho^3\cos\theta$	About tangential plane Line in any z plane about x axis	About $A_d = 0$	About $A_d = 0$	If rotated by π about z axis

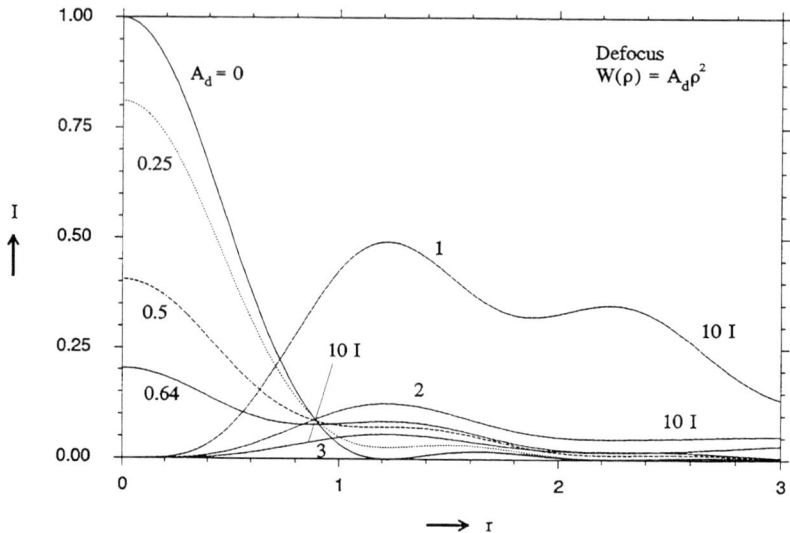

Figure 8–5. PSFs of a defocused system. A_d **represents the peak value of defocus aberration in units of** λ.

8.4.1 Defocus ($A_d\rho^2$)

Figure 8-5 shows defocused PSFs for several values of the peak defocus aberration A_d, such that the aberration-free central value is unity. As A_d increases, the central value, i.e., the Strehl ratio, decreases. The minima of the aberration-free PSF are no longer zero. The central value is zero when A_d is an integral number of waves, as may be seen from Eq. (8-7). Since the corresponding PSFs have very low values, they have been multiplied by a factor of 10 in the figure. The PSF for $A_d = 0.64\lambda$ is included here because, as will be discussed in Section 8.7, the OTF is negative for certain spatial frequency bands for larger values of A_d.

8.4.2 Spherical Aberration Combined with Defocus ($A_s\rho^4 + A_d\rho^2$)

The irradiance distribution is rotationally symmetric about the z axis. Accordingly, it is radially symmetric in any observation plane (normal to the z axis), regardless of the value of the Fresnel number. For large Fresnel numbers, the irradiance along the z axis is symmetric about the defocused point corresponding to $A_d = -A_s$. The irradiance distributions in two observation planes located symmetrically about this point are not equal to each other unless they are for aberration coefficients with opposite signs. The distribution in the $z = R$ plane does not change if we change the sign of the aberration coefficient A_s.

Figure 8-6 shows the PSFs for spherical aberration combined with defocus. Figure 8-6a shows the PSFs in defocused image planes $A_d = -A_s$ corresponding to minimum aberration variance. The PSF for $A_s = 2.2\lambda$ is also included here because, as discussed later in Section 8.7, the OTF is negative for certain spatial frequency bands for larger values of A_s. We note that the radius of the central bright spot does not change as the

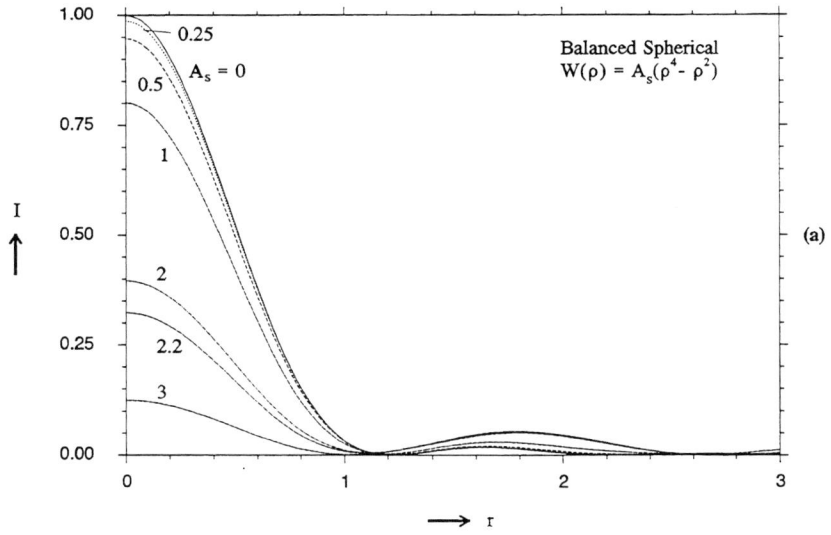

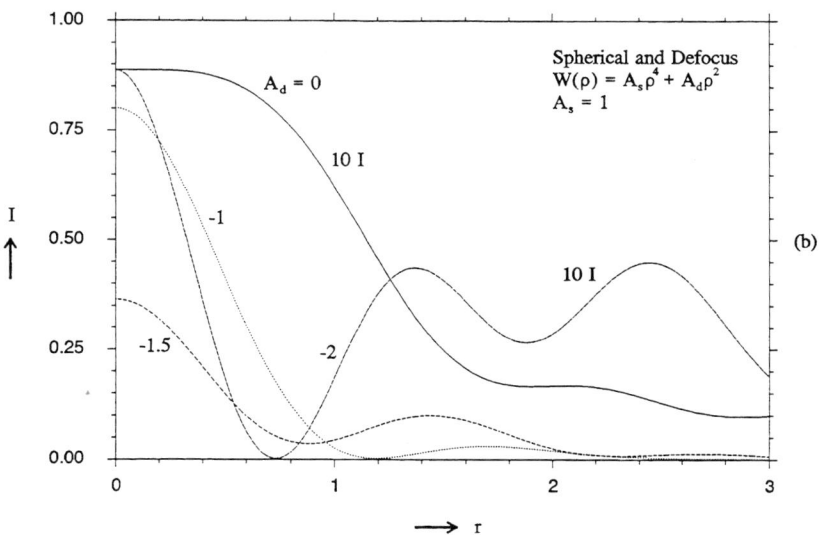

Figure 8–6. PSFs for spherical aberration. (a) Various values of A_s and defocused image planes $A_d = -A_s$ corresponding to minimum aberration variance. (b) Fixed A_s and various image planes. A_d and A_s are in units of λ.

aberration is increased.[11] Figure 8-6b shows the PSFs for $A_s = 1\lambda$ and $A_d/A_s = 0, -1$, -1.5, and -2, corresponding to paraxial, minimum-aberration-variance, circle-of-least-confusion, and marginal image planes, respectively. The PSFs in the paraxial and marginal image planes have been multiplied by a factor of 10 in this figure because of their low values. We note that the central irradiances in these two planes are equal to each other, showing that the axial irradiance is symmetric about the $A_d = -A_s$ plane, as pointed out in Table 8-3. It is also clear from the figure that the best image is obtained in the plane corresponding to minimum aberration variance.

8.4.3 Astigmatism Combined with Defocus ($A_a\rho^2\cos^2\theta + A_d\rho^2$)

The irradiance distribution for astigmatism in any observation plane is symmetric about two orthogonal axes, one of them lying in the tangential plane. It is four-fold symmetric in the plane $A_d = -A_a/2$. The distributions in two observation planes located symmetrically about the plane corresponding to $A_d = -A_a/2$ are equal to each other if we rotate one with respect to the other by $\pi/2$ or if they are for aberration coefficients with opposite signs.

Figure 8-7 shows the PSFs for various values of A_a in defocused planes $A_d = -A_a/2$, corresponding to minimum aberration variance, along the directions $\theta_i = 0$ and $\pi/4$. Figure 8-8 shows the PSFs for $A_a = 1\lambda$ and $A_d/A_a = 0, -1/2$, and -1, corresponding to paraxial or sagittal-line, minimum- aberration-variance (circle of least confusion), and tangential-line image planes, respectively. (As discussed in Section 7.5, the two line-image planes are those planes in which the astigmatic focal lines based on geometrical optics are obtained.) We note from Figure 8-8a that the central irradiance of the line images are equal to each other, showing that the axial irradiance is symmetric about the $A_d = -A_a/2$ plane as pointed out in Table 8-3. The distributions of the two line images are rotated with respect to each other by $\pi/2$. Accordingly, the solid curve in Figure 8-8b corresponds to both $A_d/A_a = 0$ and -1.

8.4.4 Coma ($A_c\rho^3\cos\theta$)

The irradiance distribution for coma is symmetric about the tangential plane. Thus, it has a line symmetry in any observation plane, the line lying in the tangential plane. The distributions in two observation planes located symmetrically about the $z = R$ plane are identical. A change in the sign of the aberration coefficient A_c produces a rotation of the distribution by π about the z axis.

Figure 8-9 shows the PSFs for coma along the directions for $\theta_i = 0, \pi/4, \pi/2$, where $\theta_i = 0$ corresponds to the x_i axis. We note that, as pointed out in Section 8.3, the peak value of an aberrated PSF does not lie at the Gaussian image point ($r = 0$). Figure 8-9c shows the symmetry of the PSFs about the x_i axis.

Pictures of the PSFs for various values of a primary aberration are shown in Figures 8-10 and 8-11. The purpose of showing these computer-generated pictures is to illustrate how the aberrated PSFs would appear in practice. The emphasis of these pictures is on the structure of a PSF, i.e., on the distribution of its bright and dark *regions*, and not on its irradiance distribution. For example, regions of very low irradiance have

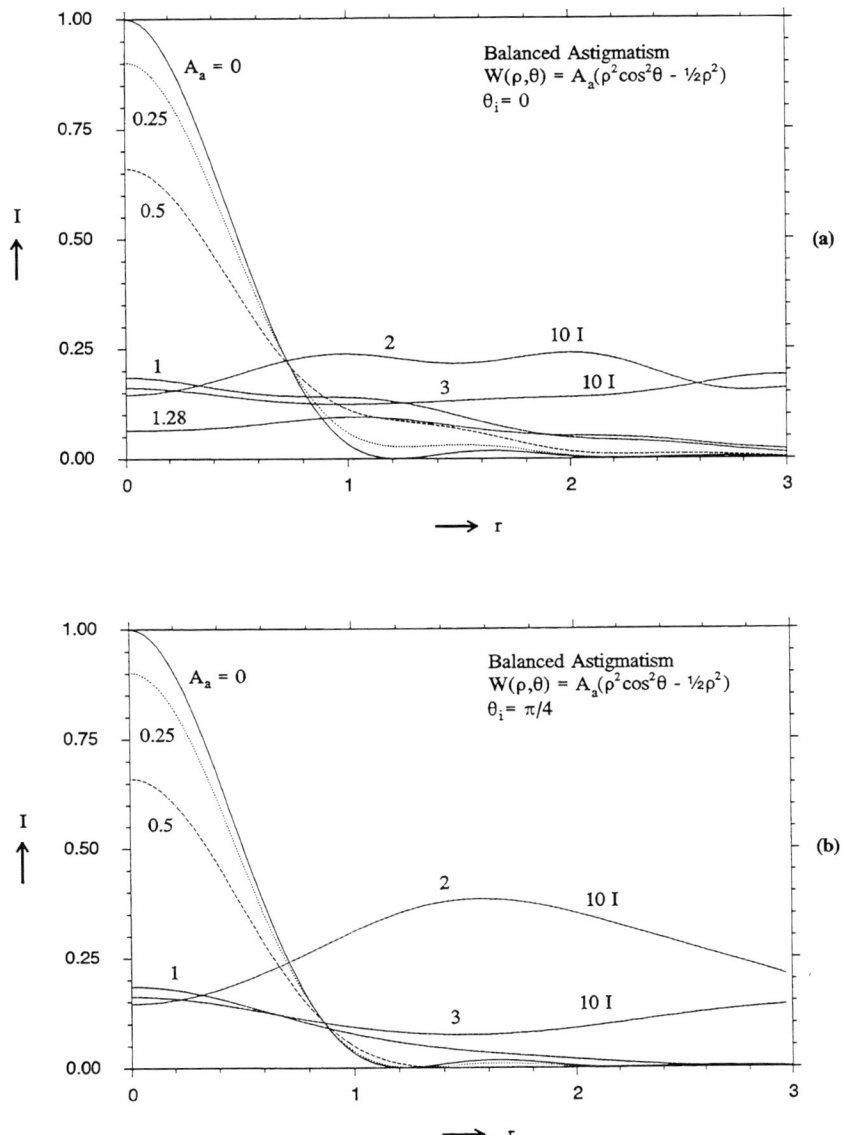

Figure 8–7. PSFs for various amounts of astigmatism in defocused image planes corresponding to minimum aberration variance along the directions $\theta_i = 0$ and $\pi/4$. A_a represents the peak value of astigmatism aberration in units of λ, and a defocused image plane is represented by $A_d = -A_a/2$.

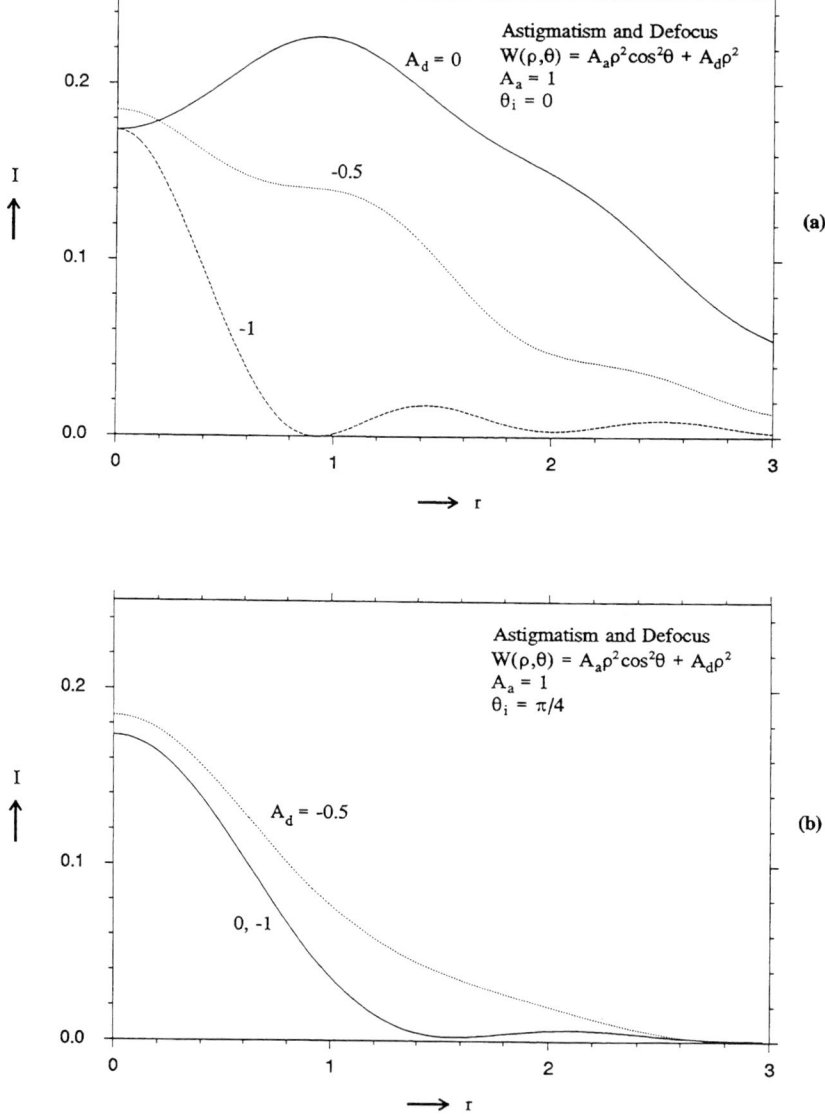

Figure 8–8. PSFs for astigmatism with $A_a = 1\lambda$ in image planes $A_d/A_a = 0, -1/2$, and -1, corresponding to sagittal-line, minimum-aberration-variance, and tangential-line image planes, respectively, along the directions $\theta_i = 0$ and $\pi/4$.

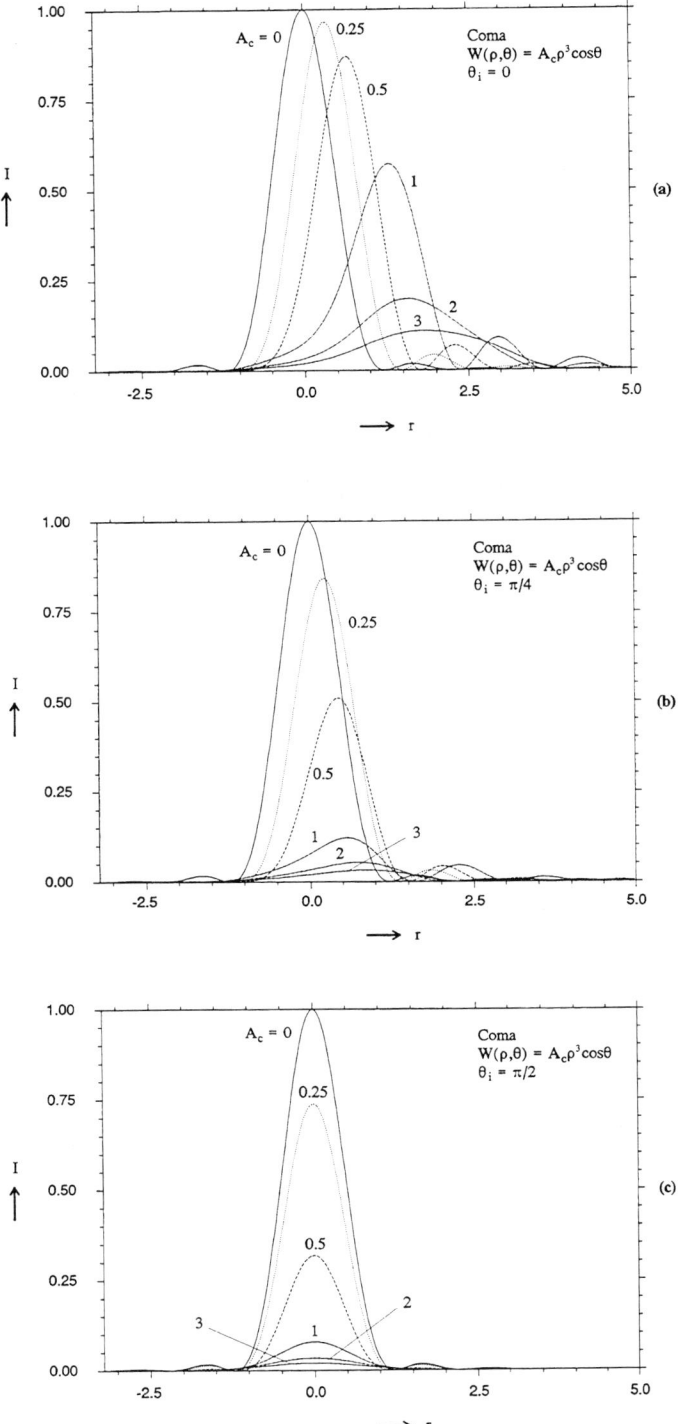

Figure 8–9. PSFs for coma along the directions (a) $\theta_i = 0$ (b) $\theta_i = \pi/4$, and (c) $\theta_i = \pi/2$. A_c is the peak value of coma aberration in units of λ.

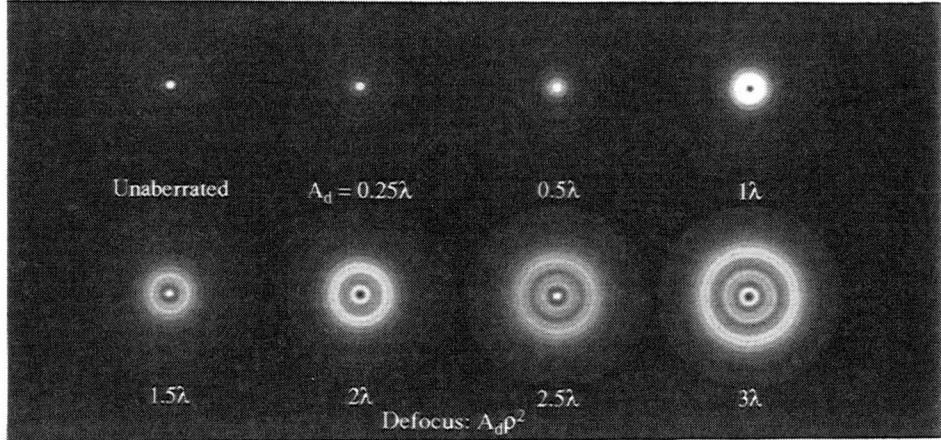

(a)

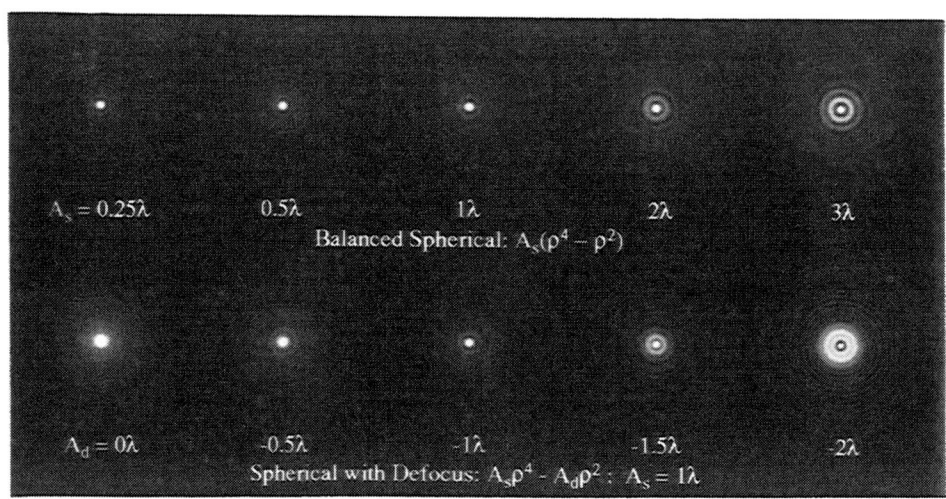

(b)

Figure 8–10. Pictures of PSFs for various amounts of (a) defocus and (b) spherical aberrations.

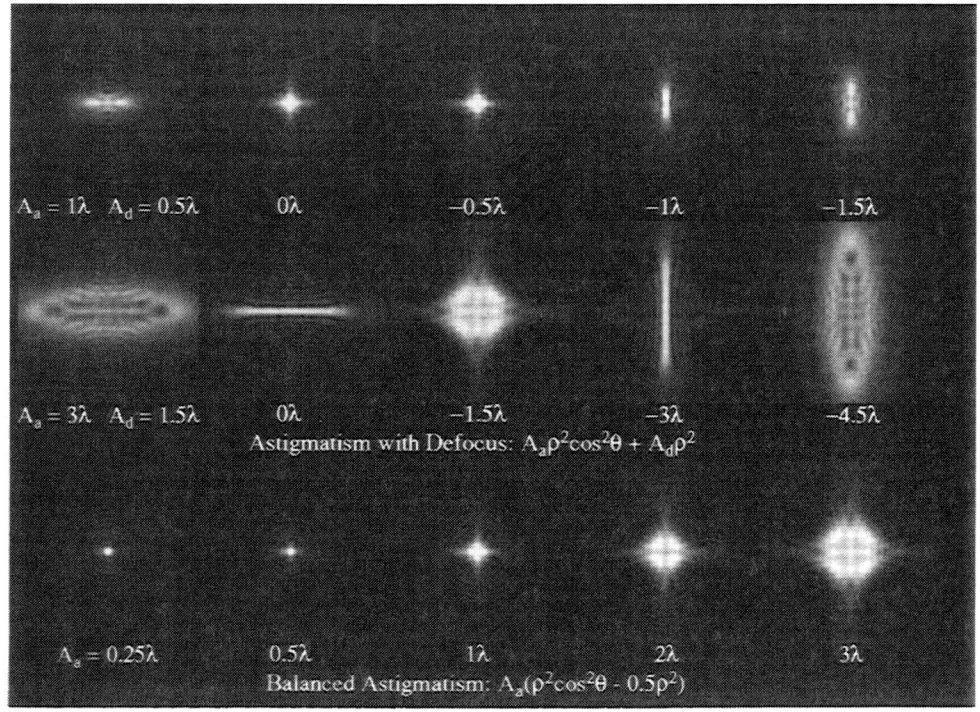

(a)

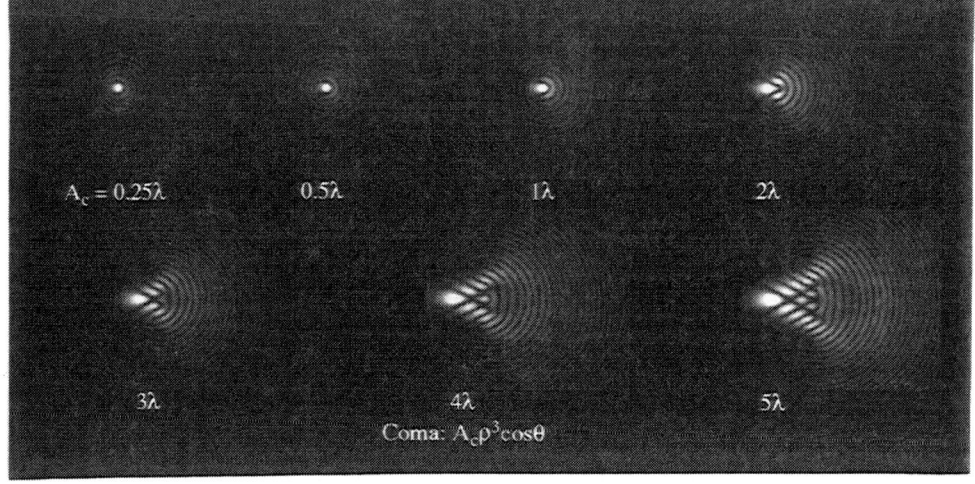

(b)

Figure 8–11. Pictures of PSFs for various amounts of (a) astigmatism and (b) coma aberrations.

been accentuated to make their appearance in these pictures. Some of the symmetry-properties of aberrated PSFs discussed above are evident from these pictures. It should be clear that a random mixture of various aberrations will only lead to a complicated PSF.

8.5   Optical Transfer Function (OTF)

Since the diffraction image of an incoherent object is given by the convolution of its Gaussian image and the system PSF, a Fourier transform of this relationship shows that the spatial frequency spectrum of the diffraction image is given by the product of the spectrum of the Gaussian image and the optical transfer function (OTF) of the system, where the OTF is equal to the Fourier transform of the PSF.[1,2] Because of the relationship of Eq. (8-1) between the PSF and the pupil function of the system, the OTF is also given by the autocorrelation of the pupil function. Thus, the OTF of a system can be obtained from its pupil function without having to calculate its PSF. In this section, we introduce the concept of OTF and discuss its physical significance. We also discuss how it is affected by aberrations and how it relates to the Strehl ratio. Also given is an expression for the aberration-free OTF of a system with a circular pupil.

The OTF of an incoherent imaging system is given by the Fourier transform of its PSF according to

$$\tau(\vec{v}_i) = \int PSF(\vec{r}_i) \exp(2\pi i \vec{v}_i \cdot \vec{r}_i) d\vec{r}_i , \qquad (8\text{-}29)$$

where $\vec{v}_i = (v_i, \phi)$ is a 2-D *spatial frequency* vector in the image plane, $\vec{r}_i = (\lambda Fr, \theta_i)$ is the position vector of a point in this plane, and the PSF is given by Eq. (8-1) with $P = 1$. In what follows, we assume that the Fresnel number of the system is large so that the defocus tolerance dictates that $z \simeq R$. However, if this is not the case, then we simply replace R by z in the following discussion. As mentioned above, because of the relationship of Eq. (8-1) between the PSF and the pupil function, the OTF may also be written as the autocorrelation of the pupil function, i.e.,

$$\tau(\vec{v}_i) = S_p^{-1} \int P(\vec{r}_p) P^*(\vec{r}_p - \lambda R \vec{v}_i) d\vec{r}_p , \qquad (8\text{-}30)$$

where

$$P(\vec{r}_p) = \exp[i\Phi_0(\vec{r}_p)], \quad 0 \le |\vec{r}_p| \le a$$
$$= 0, \quad \text{otherwise} \qquad (8\text{-}31)$$

is the pupil function. Here, $\vec{r}_p = (a\rho, \theta)$ is the position vector of a point in the plane of the exit pupil. The integration in Eq. (8-30) is carried out across the region of overlap of two pupils centered at $\vec{r}_p = 0$ and $\vec{r}_p = \lambda R \vec{v}_i$. The asterisk in Eq. (8-30) indicates a complex conjugate. We have added a subscript to the aberration function for later convenience [e.g., see Eq. (8-42), where no subscript is used].

The OTF depends on the wavelength in two ways. First, the dependence of the phase aberration on it is evident. Second, it enters in the displacement of the pupil. It has the implication that for a longer wavelength the displacement approaches the diameter of the pupil for a smaller frequency, thereby reducing the region of overlap of two pupils displaced with respect to each other to zero. Consequently, the OTF approaches zero for a smaller frequency.

The physical significance of the OTF may be understood with the help of Figure 8-12. If we consider a sinusoidal object of a spatial frequency $\vec{v}_0$, modulation or contrast m, and phase δ, its Gaussian image is also sinusoidal with a spatial frequency $\vec{v}_i = \vec{v}_0/M$, where M is the magnification of the image. The modulation and phase of the Gaussian image are the same as those of the object. Its diffraction image is also sinusoidal with a spatial frequency $\vec{v}_i$. However, the modulation of the diffraction image is $m|\tau(\vec{v}_i)|$ and its phase is $\delta - \Psi(\vec{v}_i)$, where $|\tau(\vec{v}_i)|$ is the modulus of the OTF and $\Psi(\vec{v}_i)$ is its phase, i.e.,

$$\tau(\vec{v}_i) = |\tau(v_i)| \exp[i\Psi(\vec{v}_i)] . \tag{8-32}$$

The functions $|\tau(v_i)|$ and $\Psi(\vec{v}_i)$ are also called the *modulation transfer function* (MTF) and the *phase transfer function* (PTF), respectively. We note from Eq. (8-29) that a change in the argument of the PSF from $\vec{r}_i$ to $\vec{r}_i{}'$ introduces a phase shift of $2\pi\vec{v}_i \bullet (\vec{r}_i - \vec{r}_i{}')$ in the OTF varying linearly with the spatial frequency $\vec{v}_i$. In other words, a linear component of the PTF corresponds to a shift of the image as a whole.

From Eq. (8-30), we note that $\tau(0) = 1$; i.e., the image of a uniform (zero spatial-frequency) object is also uniform regardless of the aberration in the system. From Eq.

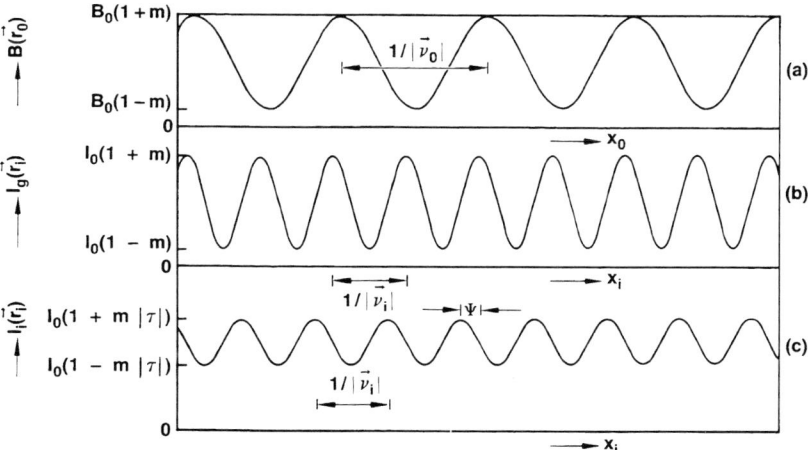

Figure 8–12. Image of a sinusoidal object. (a) Object of spatial frequency v_0 and modulation m. (b) Gaussian image of spatial frequency $v_i = v_0/M$ and modulation m. (c) Diffraction image of spatial frequency v_i, modulation $m|\tau|$ and phase Ψ. B_0 and I_0 represent the average radiance and irradiance of the sinusoidal object and image, respectively.

(8-29), we note that $\tau(\vec{v}_i) = \tau * (-\vec{v}_i)$; i.e., the OTF is complex symmetric or Hermitian. Thus, the real part of the OTF is an even function and its imaginary part is an odd function of the spatial frequency. Similarly, the MTF is an even function and the PTF is an odd function of the spatial frequency. Accordingly, the real parts of the OTFs, or the MTFs for spatial frequencies (v_i, ϕ) and $(v_i, \pi + \phi)$, are equal. However, their imaginary parts, or the PTFs for these frequencies, have equal magnitudes but opposite signs. It can be shown that $|\tau(\vec{v}_i)| \leq 1$; i.e., the MTF at any spatial frequency is less than or equal to 1. The aberrated MTF at any spatial frequency is less than or equal to the corresponding aberration-free MTF. Moreover, the slope of the real part of the OTF at the origin is equal to the corresponding slope of the MTF. This slope is independent of the aberration in the system.

For an aberration-free system with a circular pupil, Eq. (8-30) for its OTF reduces to

$$\tau(v) = (2/\pi)[\cos^{-1}v - v(1-v^2)^{1/2}], 0 \leq v \leq 1$$

$$= 0, \text{ otherwise} \tag{8-33}$$

where $v = \lambda F v_i$ is a normalized radial spatial frequency. The OTF in this case corresponding to a spatial frequency $\vec{v}_i$ represents the fractional area of overlap between two circles whose centers are separated by a distance $\lambda R v_i$. We note that the OTF is radially symmetric; i.e.; its value depends on the magnitude of a spatial frequency but not on its direction. The frequency $v_i = 1/\lambda F$, corresponding to $v = 1$, is called the *cutoff frequency* of the system. For example, a system with a focal ratio $F = 10$ imaging an object radiating at a wavelength $\lambda = 0.5$ μm corresponds to a cutoff frequency of 200 cycles/mm. The cutoff frequency decreases linearly with wavelength. The sinusoidal components of an object with spatial frequencies $v_0 \geq M/\lambda F$ are not resolved by the system at all; i.e., their images are of uniform irradiance. From Eq. (8-33), we find that

$$\tau'(0) = \left[\frac{\partial \tau(v)}{\partial v} \right]_{v=0} \tag{8-34a}$$

$$= -4/\pi, \tag{8-34b}$$

and

$$\int_0^1 \tau(v)v\,dv = 1/8. \tag{8-35}$$

Since the slope of the MTF or the real part of the OTF of a system evaluated at the origin is independent of its aberration, it is equal to $-4/\pi$ in the case of a circular pupil regardless of its aberration.

The Strehl ratio of an optical imaging system, discussed in Section 8.3, represents the ratio of its PSF (or the corresponding irradiance) values at the center $r = 0$ with and without aberration. From Eq. (8-29), we note that its PSF can be written as the inverse Fourier transform of its OTF, i.e.,

$$PSF(\vec{r}_i) = \int \tau(\vec{v}_i) \exp(-2\pi i \vec{v}_i \cdot \vec{r}_i) d\vec{v}_i \ . \tag{8-36}$$

Accordingly, the Strehl ratio may be written

$$S = (4/\pi) \int \tau(\vec{v}) d\vec{v} , \tag{8-37}$$

where we have used Eq. (8-35) for the integral involving the aberration-free OTF. Since S is a real quantity, the integral of the imaginary part of $\tau(\vec{v})$ on the right-hand side of Eq. (8-37) must be zero. Hence, we may write Eq. (8-37) as

$$S = (4/\pi) \int Re\tau(\vec{v}) d\vec{v} , \tag{8-38}$$

where Re indicates a real part. Thus, the Strehl ratio of a system gives a measure of the mean value of the real part of its OTF, averaged over all spatial frequencies.

8.6   Hopkins Ratio

In Section 8.3, we calculated aberration tolerances for a system with a Strehl ratio of 0.8. Such a system forms the image of an object with a quality that is only slightly inferior to the corresponding quality for an aberration-free system, regardless of the spatial frequencies (or the size of detail) of interest in the object. A Strehl ratio of 0.8 is obtained when the standard deviation of the aberration of the system across its exit pupil is approximately $\lambda/14$, regardless of the type of aberration. However, systems having much larger aberrations form good-quality images of objects in which the size of the detail is much coarser than the limiting resolution $1/\lambda F$ of the system.

We now consider aberration tolerances based on a certain amount of reduction in the MTF of the system corresponding to a certain spatial frequency. Following Hopkins,[12] we define a modulation ratio $H(\vec{v}_i)$ as the ratio of the MTFs of a system at a spatial frequency $\vec{v}_i$ with and without aberration, i.e.,

$$H(\vec{v}_i) = |\tau(\vec{v}_i)| / \tau_o(v_i) , \tag{8-39}$$

where $\tau(\vec{v}_i)$ is the aberrated OTF, and $\tau_0(v_i)$ is the aberration-free OTF given by Eq. (8-33) with v replaced by v_i and $0 \le v_i \le 1/\lambda F$. For obvious reasons, we call $H(\vec{v}_i)$ the *Hopkins modulation* or *contrast ratio*. In terms of the pupil function, it may be written

$$H(\vec{v}_i) = \frac{1}{\tau_0(v_i)S_p} \left| \int \exp\{i[\Phi_0(\vec{r}_p) - \Phi_0(\vec{r}_p - \lambda R\vec{v}_i)]\} d\vec{r}_p \right| \ . \tag{8-40}$$

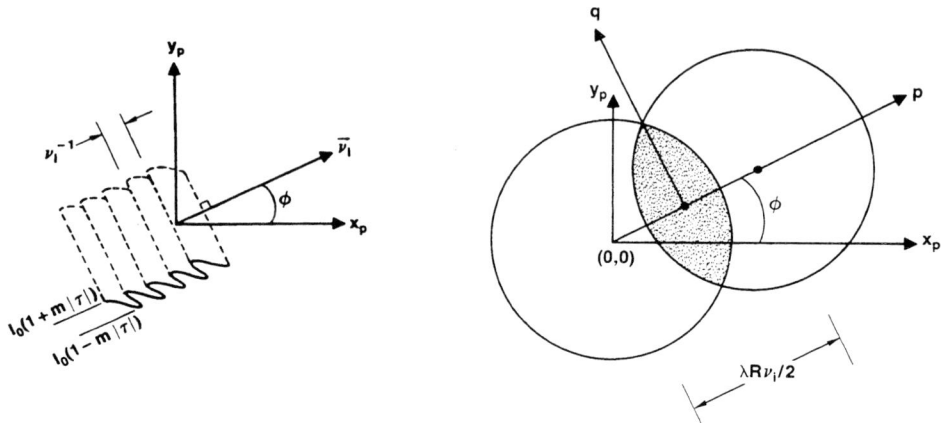

Figure 8–13. Geometry for evaluating the OTF of a system. Two pupils are shown, one centered at the origin and the other centered at $\lambda R\vec{v}_i$ in the (x_p,y_p) coordinate system. The (p,q) coordinate system is rotated with respect to the (x_p,y_p) system by an angle ϕ; i.e., the p axis lies along the $\vec{v}_i$ direction. Its origin lies at the midpoint of the line joining the centers of the two pupils. Thus, in the (p,q) coordinate system, the pupils are centered at $\pm (\lambda R/2)(v_i,0)$.

If we consider the aberration function $\Phi(\vec{r}_p)$ in a rotated coordinate system (p,q) shown in Figure 8-13 such that the p axis lies along the direction of $\vec{v}_i$ and its origin lies at a distance of $\lambda R v_i/2$ along the p axis from the origin of the (x_p,y_p) coordinate system, then Eq. (8-40) may be written

$$H(\vec{v}_i) = \frac{1}{\tau_0(v_i)S_p} \left| \int\int \exp[iQ(p,q,\vec{v}_i)]dpdq \right| ,$$
(8-41)

where

$$Q(p,q;\vec{v}_i) = \Phi(p + \lambda Rv_i/2,q) - \Phi(p - \lambda Rv_i/2,q)$$
(8-42)

is an *aberration difference function*. The aberration function $\Phi(p,q)$ is obtained from $\Phi_0(x_p,y_p)$ according to

$$\Phi(p,q) = \Phi_0(p\cos\phi - q\sin\phi, p\sin\phi + q\cos\phi) .$$
(8-43)

As discussed in Section 1.6, the aberration function $\Phi_0(x_p,y_p)$ for a rotationally symmetric system depends on x_p and y_p through $x_p^2 + y_p^2$ and x_p, where $x_p = r_p\cos\theta$, $y_p = r_p\sin\theta$, and θ is the angle between the vector $\vec{r}_p$ and the x_p axis which lies in the tangential plane. Hence $\Phi(p,q)$ is obtained from $\Phi_0(x_p,y_p)$ by replacing $x_p^2 + y_p^2$ by $p^2 + q^2$ and x_p by $p\cos\phi - q\sin\phi$.

Equation (8-41) may also be written

$$H(\vec{v}_i) = |<\exp\{i[Q-<Q>\,]\}>|\ , \tag{8-44}$$

where the angular brackets indicate an average. For example,

$$<Q> = \frac{1}{\tau_0(v_i)S_p}\int\int Q(p,q;\vec{v}_i)dpdq\ . \tag{8-45}$$

For small values of $Q - <Q>$, we may retain only the first three terms in the expansion of the exponential, and thus obtain the result

$$H(\vec{v}_i) \simeq 1 - \frac{1}{2}\sigma_Q^2\ , \tag{8-46}$$

where

$$\sigma_Q^2 = <Q^2> - <Q>^2 \tag{8-47}$$

is the variance of the aberration-difference function across the overlap region of the two pupils. If, in place of Eq. (8-39), we consider

$$H(\vec{v}_i)\exp[i\Psi(\vec{v}_i)] = \tau(\vec{v}_i)/\tau_0(v_i) \tag{8-48}$$

we find that

$$\Psi(\vec{v}_i) = <Q>\ ; \tag{8-49}$$

i.e., $<Q>$ represents the phase transfer function.

We noted in Section 8.3.1 that the Strehl ratio of an aberrated system depends on the variance of the aberration function across its pupil and not on the type of the aberration. Similarly, we note from Eq. (8–46) that the Hopkins ratio at a certain spatial frequency depends on the variance of the aberration differene function across the overlap region of two displaced pupils (displacement depending on the spatial frequency) and not on the type of the aberration. As in the case of Strehl ratio, a better approximation to Hopkins ratio is obtained by using the exponential relation

$$H(\vec{v}_i) = \exp\left(-\frac{1}{2}\sigma_Q^2\right)\ . \tag{8-50}$$

Szapiel[13] has shown that for primary aberrations, when the true value of $H(\vec{\nu}_i)$ exceeds 0.4, the approximate realtion of Eq. (8–50) estimates it to within 0.06.

Based on numerical analysis, Hopkins[12] has shown that $H(\nu) \geq 0.8$ for $\nu \leq 0.1$, provided the primary aberration coefficients obey the following conditions:

$$A_d \leq \pm \lambda/20\nu \tag{8-51}$$

$$A_a \leq \pm \lambda/10\nu \text{ in the plane } A_d = -A_a/2 \tag{8-52}$$

$$A_c \leq \pm \lambda\left(\frac{0.071}{\nu} + 0.16\right) \text{ with } \Psi(\nu) = \mp 0.89 + 0.48\nu \text{ when } \phi = \pi/2 \tag{8-53a}$$

$$\leq \pm \lambda\left(\frac{0.123}{\nu} + 0.19\right) \text{ with } \Psi(\nu) = 0 \text{ when } \phi = 0 \tag{8-53b}$$

$$A_s \leq \pm \lambda\left(\frac{0.106}{\nu} + 0.33\right) \text{ in the plane } A_d = -(1.33 - 2.2\nu + 2.8\nu^2)A_s \ . \tag{8-54}$$

As in Eq. (1-7), A_d, A_a, A_c, and A_s represent the peak coefficients of defocus, astigmatism, coma, and spherical aberration. We note that the amount of balancing defocus in the case of spherical aberration is different from its corresponding value given in Table 8-2 for optimizing the Strehl ratio. Moreover, its value depends on the magnitude of the spatial frequency at which the MTF is optimized. For spatial frequencies $\nu > 0.1$, it is more appropriate to use the Strehl ratio as the criterion of image quality and aberration balancing.

We have noted in Section 8.3 that a maximum possible value of the Strehl ratio of a system for a particular aberration is obtained by a minimization of its variance across the exit pupil, provided the Strehl ratio thus obtained is not too small. The approximate Eq. (8-46) shows that the Hopkins ratio of a system at a particular spatial frequency is maximum when the variance of the aberration difference function across the overlap region of two displaced exit pupils is minimum, the displacement of the pupils being dependent on the spatial frequency under consideration. An investigation[14] of balancing of secondary spherical aberration with primary spherical aberration and defocus, and secondary coma with primary coma, shows that unless $H(\vec{\nu}) \geq 0.7$, a decrease in the value of σ_Q may be associated with a decrease in the value of $H(\vec{\nu})$ for $\nu \leq 0.1$. Thus, the variance of the aberration or that of the aberration difference function is a useful criterion of image quality provided the corresponding Strehl or Hopkins ratio obtained is relatively high.

8.7   OTFs for Primary Aberrations

Now we discuss the full OTFs and their symmetry properties for primary aberrations. The MTFs, PTFs, and real and imaginary parts of the OTFs are considered for various values of a primary aberration. The phenomenon of *contrast reversal* is also discussed.

In the (p,q) coordinate system, Eq. (8-30) for the OTF may be written

$$\tau(\vec{v}_i) = S_p^{-1} \int \int \exp[iQ(p,q;\vec{v}_i)] \, dp \, dq \qquad (8\text{-}55a)$$

$$= S_p^{-1} \left[\int \int \cos Q \, dp \, dq + i \int \int \sin Q \, dp \, dq \right] . \qquad (8\text{-}55b)$$

We note from Figure 8-13 that the overlap region, which forms the region of integration in the above integrals, is symmetric in p and q. Hence, if Q is an odd function of p and/or q, the imaginary part of the integral vanishes. This is true for defocus, spherical aberration, and astigmatism. It is also true for coma for a spatial frequency parallel to the y_p axis, i.e., for $\phi = \pi/2$. In such cases the OTF is real and, depending on whether its value for a certain spatial frequency is positive or negative, its phase, the PTF, for that frequency is zero or π, respectively. A phase of π is sometimes associated with a negative value of the MTF. It represents *contrast reversal*, i.e., for example, bright regions in the object appear as dark regions in the image.

The aberration difference function Q for primary aberrations is given by

$$Q(p, q; v) = \begin{cases} 4A_d\, pv & \text{Defocus} \qquad (8\text{-}56a) \\[2em] 8A_s\, pv(p^2 + q^2 + v^2) + 4A_d pv & \begin{array}{l}\text{Spherical combined} \\ \text{with defocus} \qquad (8\text{-}56b)\end{array} \\[2em] 4A_a v\, (p\cos^2\phi - q\sin\phi\cos\phi) + 4A_d\, pv & \begin{array}{l}\text{Astigmatism combined} \\ \text{with defocus} \qquad (8\text{-}56c)\end{array} \\[2em] A_c(6p^2 + 2q^2 + 2v^2)\cos\phi - 4pqv\sin\phi & \text{Coma ,} \qquad (8\text{-}56d) \end{cases}$$

where (p,q) are now normalized by the pupil radius so that $p^2 + q^2 \leq 1$.

Figure 8-14 shows how the OTF of a defocused system varies with the spatial frequency. We note that it is real and radially symmetric; i.e., its value depends on the value of v but not on the value of ϕ. For $A_d \leq 0.64\lambda$, the OTF is positive for all spatial frequencies. However, for larger values of A_d, it becomes negative, corresponding to a PTF of π, for certain bands of spatial frequencies. It becomes negative for smaller and smaller spatial frequencies as the amount of defocus A_d increases. The OTF is independent of the sign of A_d (when the Fresnel number of the system is large).

To illustrate the significance of the OTF and, in particular, the contrast reversal, we consider, as shown in Figure 8-15a, a 2-D object which is sinusoidal along the vertical axis with a spatial frequency that increases linearly in the horizontal direction. The maximum frequency in the object is chosen to equal the cutoff frequency of the aberration-free system. This frequency is normalized to unity. The aberration-free or the diffraction-limited image of the object is shown in Figure 8-15b. The monotonic reduction in contrast with increasing spatial frequency is quite evident from this figure. A defocused image corresponding to $A_d = 2\lambda$ is shown in Figure 8-15c. It is clear that the contrast in the image reduces with frequency rapidly to zero, reverses its sign back and forth as the frequency increases, with practically zero values for frequencies $\nu \geq 0.3$. As a convenience, the aberration-free and defocused OTFs are shown in Figure 8-15d to illustrate the regions of zero and near zero contrast as well as the regions of contrast reversal.

Figure 8-16 shows how the OTF of a system aberrated by spherical aberration varies with the spatial frequency. As in the case of defocus, it is real and radially symmetric. Figure 8-16a shows the OTF for a defocused image plane $A_d = -A_s$, corresponding to minimum aberration variance across the pupil, for various values of A_s. We note that the OTF is positive for all spatial frequencies for $A_s \leq 2.2\lambda$. For larger values of A_s, it becomes negative for certain bands of frequencies. Figure 8-16b shows the OTF for $A_s = 1\lambda$ in various image planes. As noted in Section 7.3, $A_d/A_s = 0$, -1, -1.5, and -2 correspond to images observed in the paraxial, minimum-aberration-variance, circle-of-least-confusion, and marginal image planes, respectively. We note that, except at very high spatial frequencies, the OTF values in the minimum-aberration-variance plane are higher than the corresponding values in the other planes considered.

Figures 8-17 and 8-18 show the OTF of a system aberrated by astigmatism. Figure 8-17 shows it for a defocused image plane $A_d = -A_a/2$, corrsponding to minimum aberration variance across the pupil, for various values of A_a. The OTF in this case is real and four-fold symmetric with one axis lying in the tangential plane. We note from the figure that the OTF becomes negative for smaller values of A_a when $\phi = \pi/4$ compared to when $\phi = 0$. For example, the OTF for $\phi = 0$ is positive for $A_a \geq 1.28\lambda$.

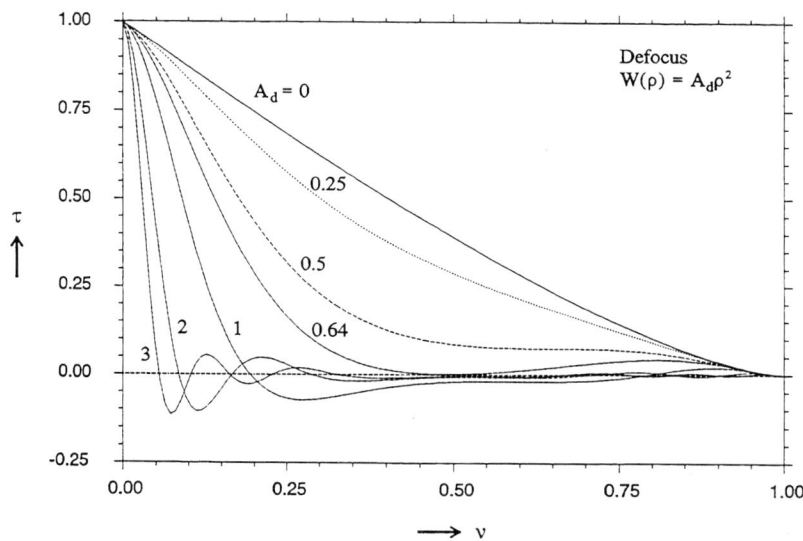

Figure 8–14. OTFs of a defocused system. A_d represents the peak defocus aberration in units of λ.

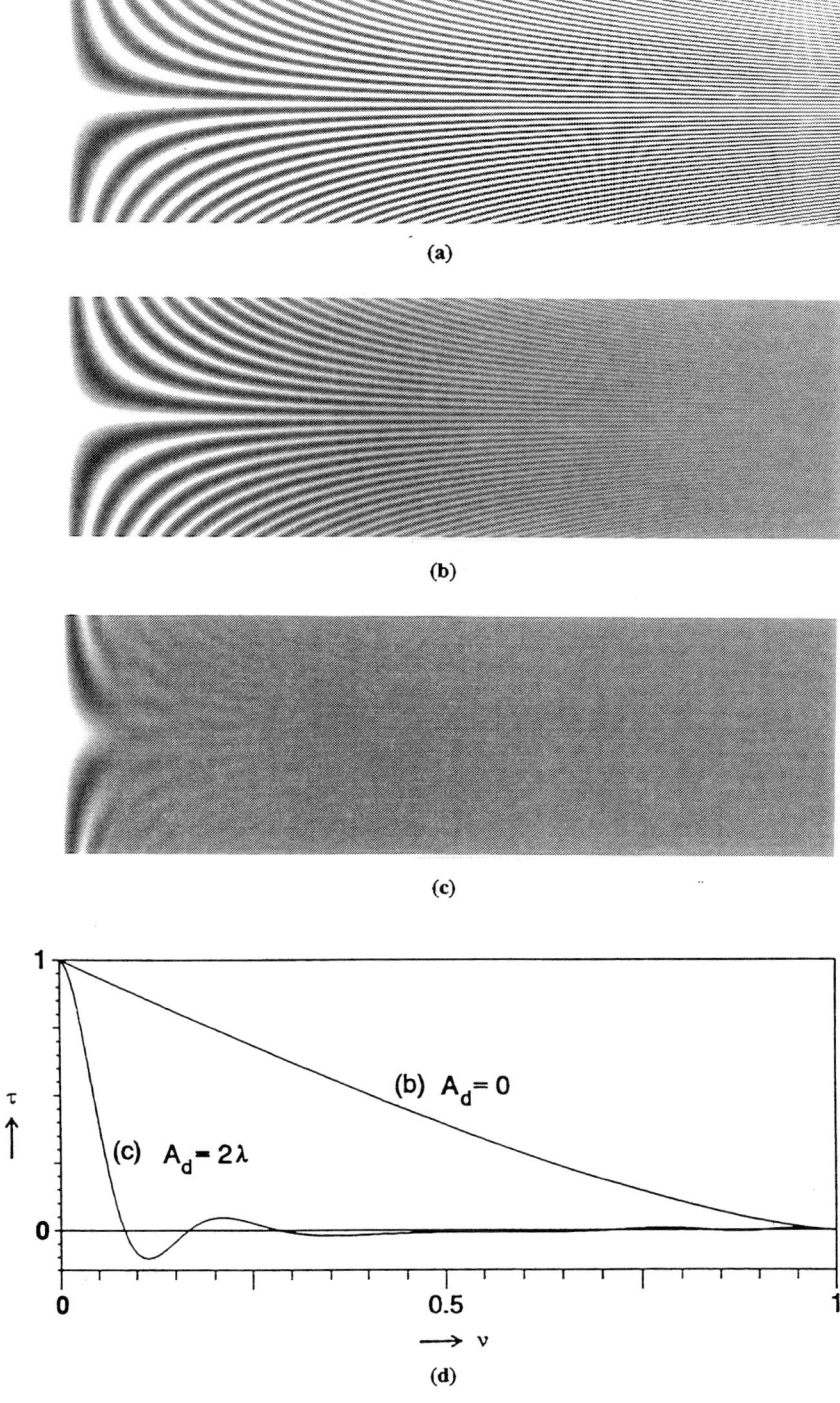

Figure 8-15. Aberration-free and defocused images of an object. (a) Object. It is sinu-soidal along the vertical axis with a spatial frequency that increases linearly in the hor-izontal direction. (b) Aberration-free image. (c) Defocused image with $A_d = 2\lambda$. (d) Aberration-free and defocused OTFs.

However, it is negative for $\phi = \pi/4$ when $A_a = 1\lambda$ and $\nu > 0.35$. Figure 8-18 shows the OTF for $A_a = 1\lambda$ in various image planes. As discussed in Section 7.5, $A_d/A_a = 0, -1/2$, and -1 correspond to image planes where the tangential focal line, circle of least confusion, and sagittal focal line are observed. The OTF is real with a biaxial (or inversion) symmetry, except for observations in the plane of least confusion, in which case it has a four-fold symmetry. The PTF is either 0 or π depending on the spatial frequency. For $\phi = 0$, the OTF in the image plane $A_d = -A_a$ is the same as the aberration-free OTF. Similarly, the OTF for $A_d = 0$ is the same as the defocused OTF in Figure 8-14 for $A_d = 1\lambda$. For $\phi = \pi/4$, the OTFs in the image planes corresponding to $A_d/A_a = 0$ and -1 are equal. If we consider the OTF for $\phi = \pi/2$, we obtain the same results for $A_d/A_a = 0, -1/2$, and -1 as those in Figure 8-18a for $A_d/A_a = -1, -1/2$, and 0, respectively.

Figures 8-19 through 8-22 show the OTF of a system aberrated by coma. In this case, the OTF is a complex function, except for $\phi = \pi/2$, i.e., when the radiance of the sinusoidal object varies along a direction that is normal to the axis about which the PSF is symmetric. Thus, the OTF dependence on the spatial frequency is shown in terms of its MTF and PTF in Figures 8-19 and 8-20, respectively. Figure 8-19 shows the MTF for several values of peak coma aberration A_c when $\phi = 0, \pi/4$, and $\pi/2$. The corresponding PTF is shown in Figure 8-20, except when $\phi = \pi/2$, in which case the PTF is zero. The value of Ψ in this figure is in radians. Note that if a wavefront tilt is added to the coma aberration to minimize its variance across the pupil as discussed in Section 8.3, it only introduces a PTF varying linearly with the spatial frequency. The real and imaginary parts $|\tau(\nu)|\cos\Psi(\nu)$ and $|\tau(\nu)|\sin\Psi(\nu)$, respectively, of the OTF are shown in Figures 8-21 and 8-22 for $\phi = 0$ and $\pi/4$. As discussed in Section 8.4, the Strehl ratio is obtained by integrating the real part, and the integral of the imaginary part is zero.

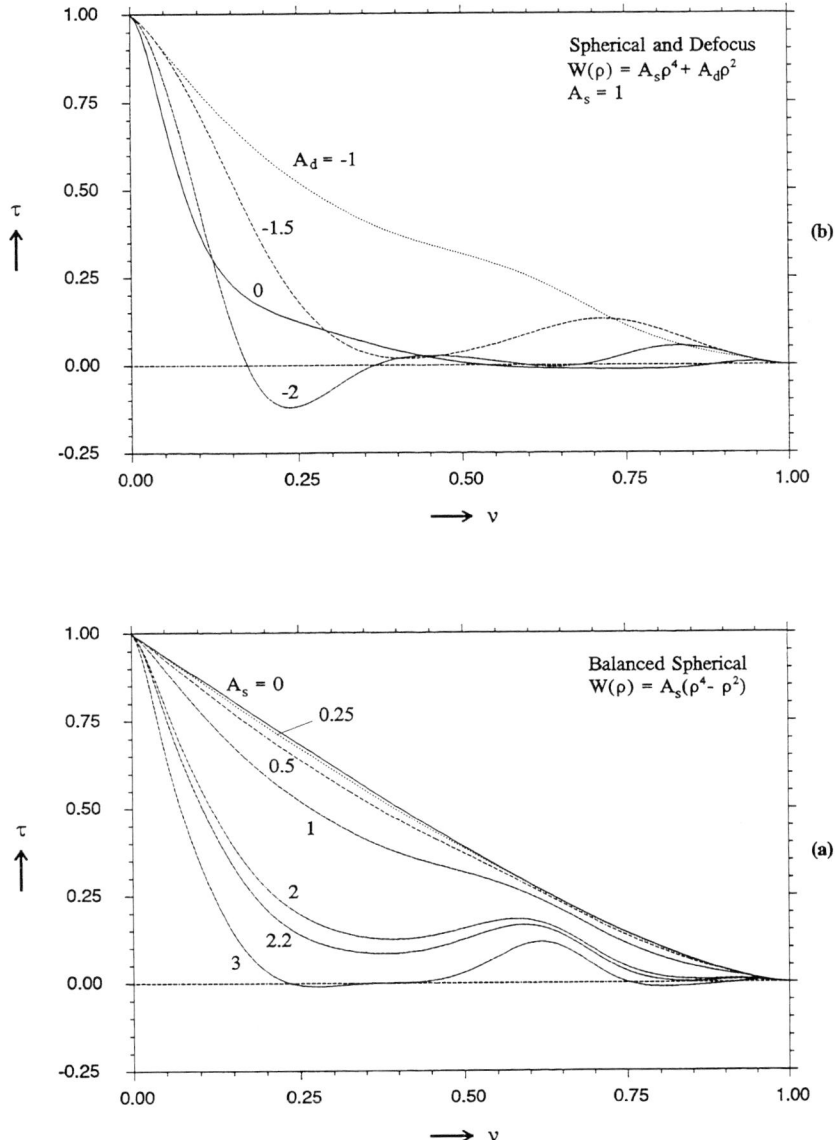

Figure 8–16. OTFs for spherical aberration. (a) Various values of A_s and image planes $A_d = -A_s$ corresponding to minimum aberration variance. (b) Fixed A_s and various image planes. A_s and A_d are in units of λ.

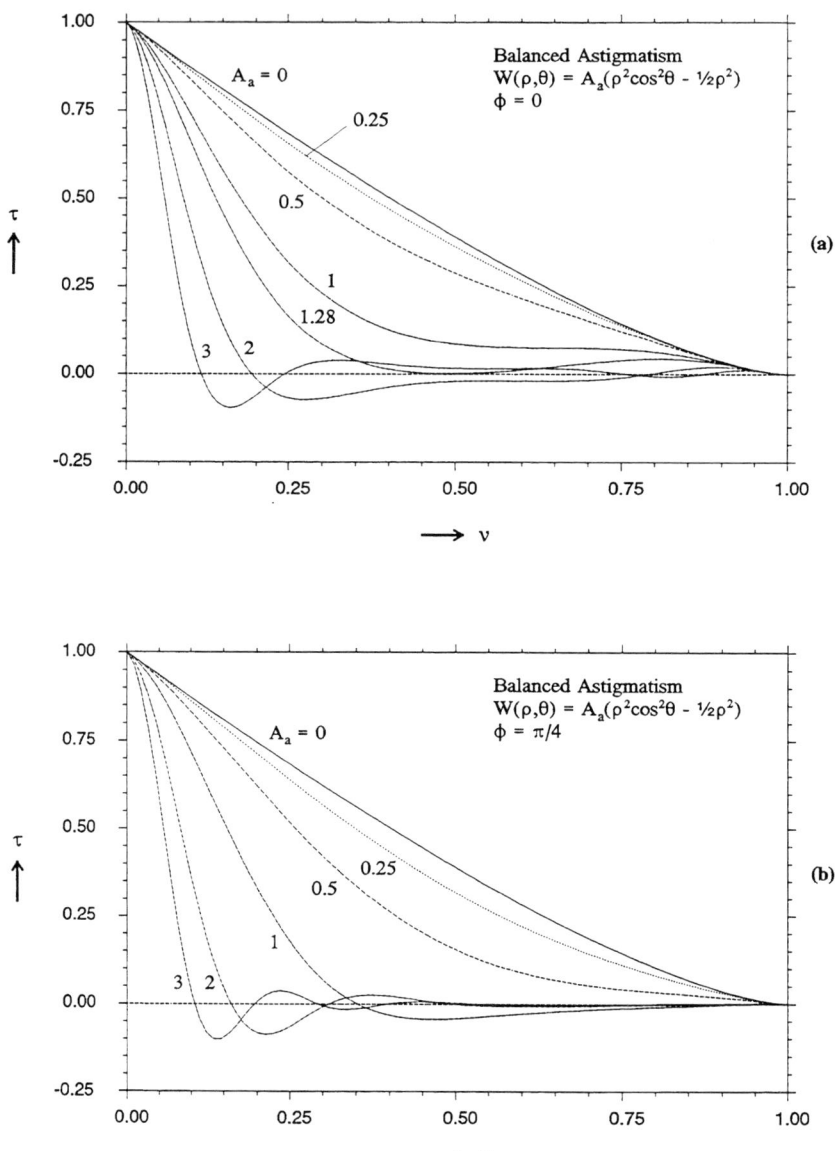

Figure 8–17. OTFs for various amounts of astigmatism in defocused image planes corresponding to minimum aberration variance. As indicated in Figure 8-13, ϕ represents the angle the spatial frequency vector makes with the x axis. A_a represents the peak value of astigmatism aberration in units of λ and a defocused image plane is represented by $A_d = -A_a/2$

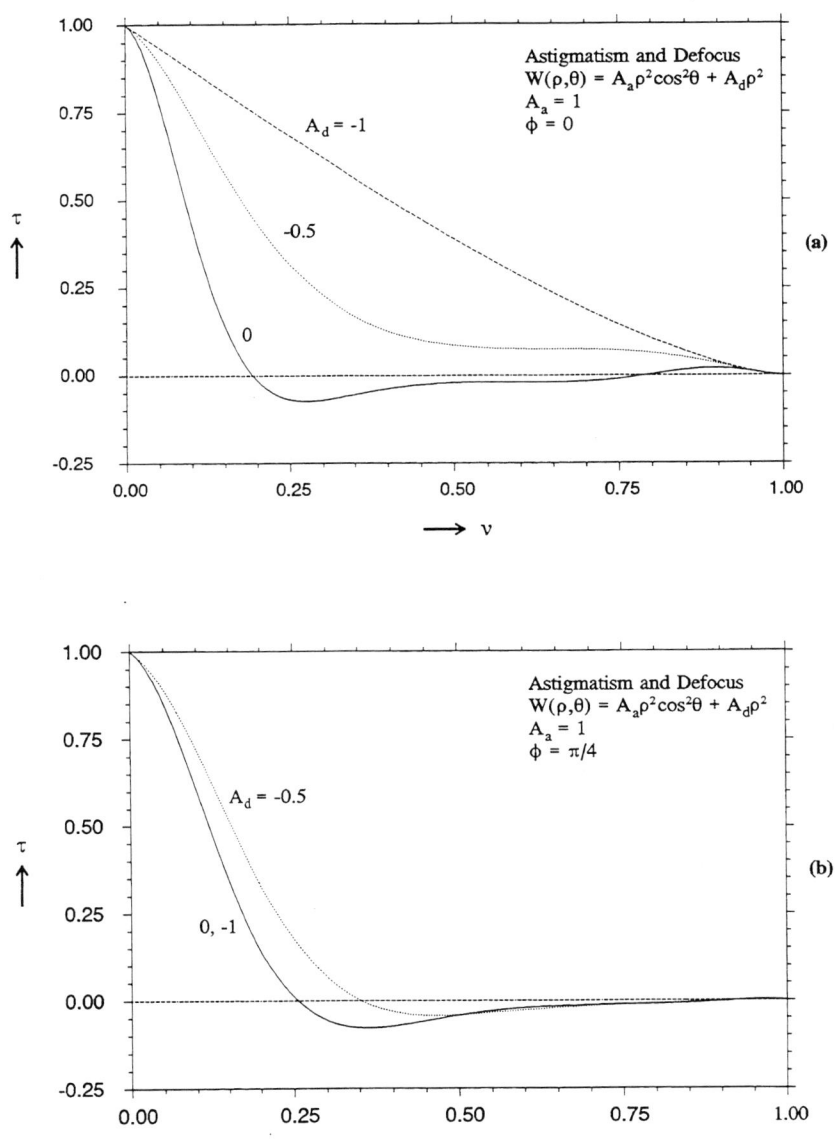

Figure 8–18. OTFs for astigmatism with $A_a = 1\lambda$ in image planes $A_d/A_a = 0, -1/2$, and -1, corresponding to sagittal-line, minimum aberration-variance, and tangential-line image planes, respectively. (a) $\phi = 0$, (b) $\phi = \pi/4$.

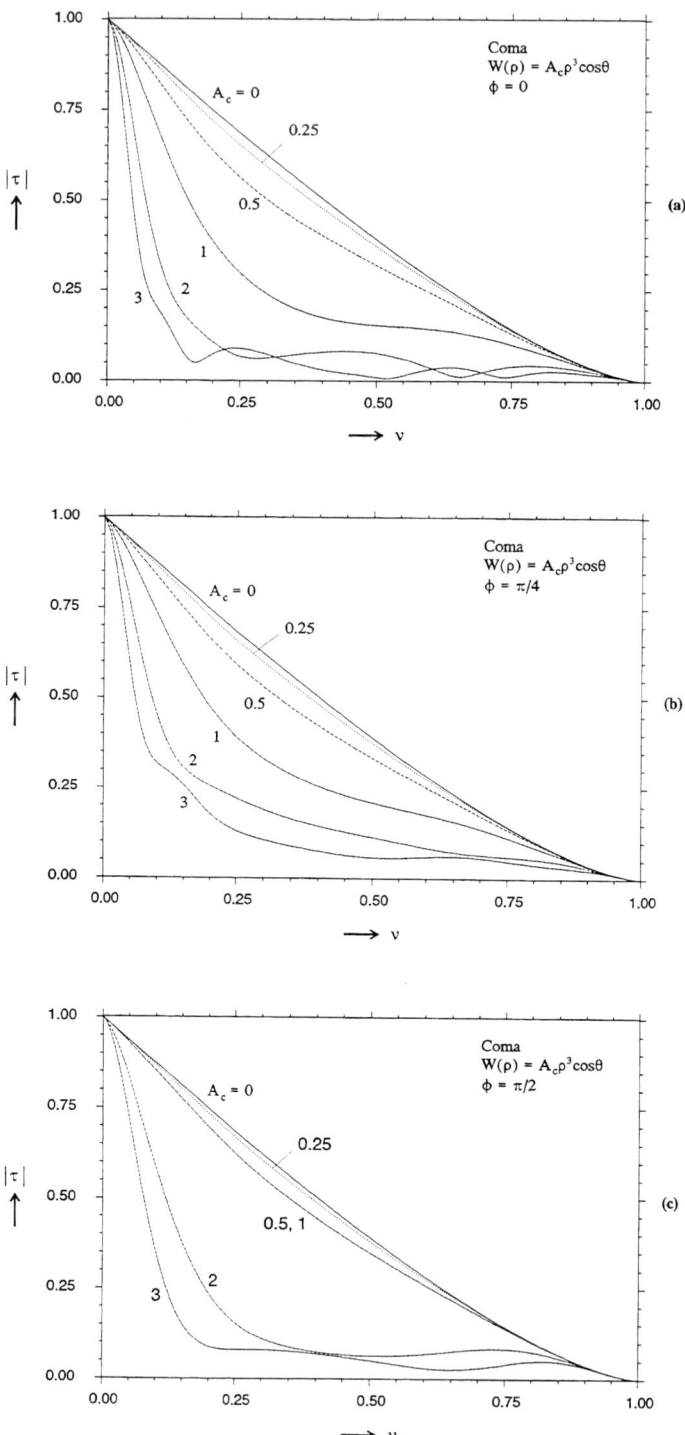

Figure 8–19. MTFs for coma. A_c **represents the peak value of coma aberration in units of** λ. (a) $\phi = 0$, (b) $\phi = \pi/4$, **and** (c) $\phi = \pi/2$.

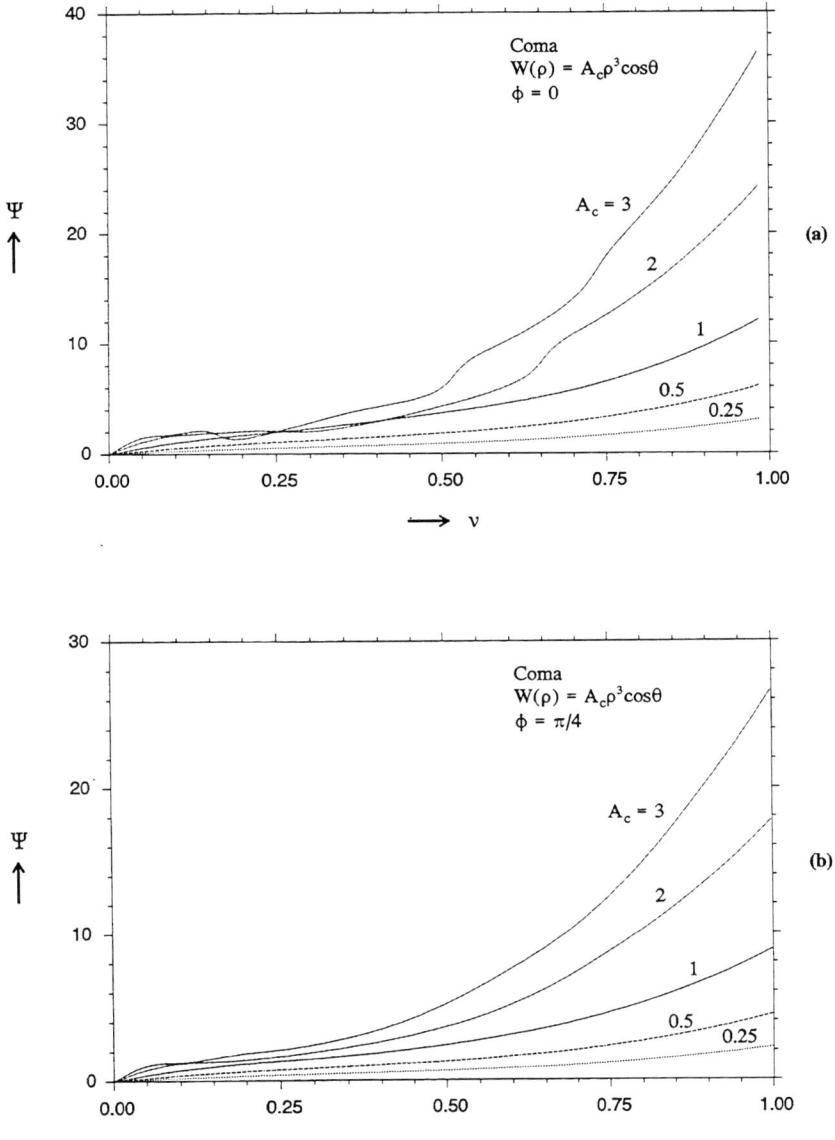

Figure 8–20. PTFs for coma. A_c **represents the peak value of coma aberration in units of** λ**. (a)** $\phi = 0$ **and (b)** $\phi = \pi/4$**.**

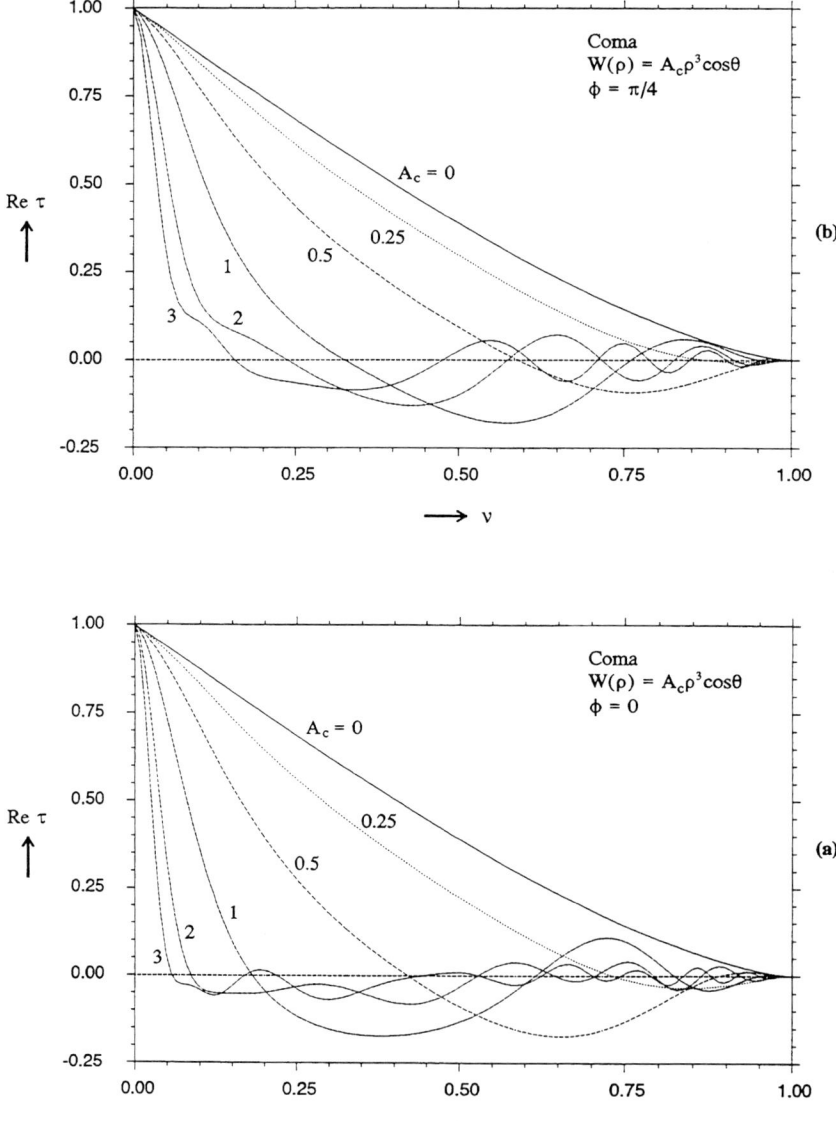

**Figure 8–21. Real part of the OTF of a system aberrated by coma.   (a) $\phi = 0$,
(b) $\phi = \pi/4$.**

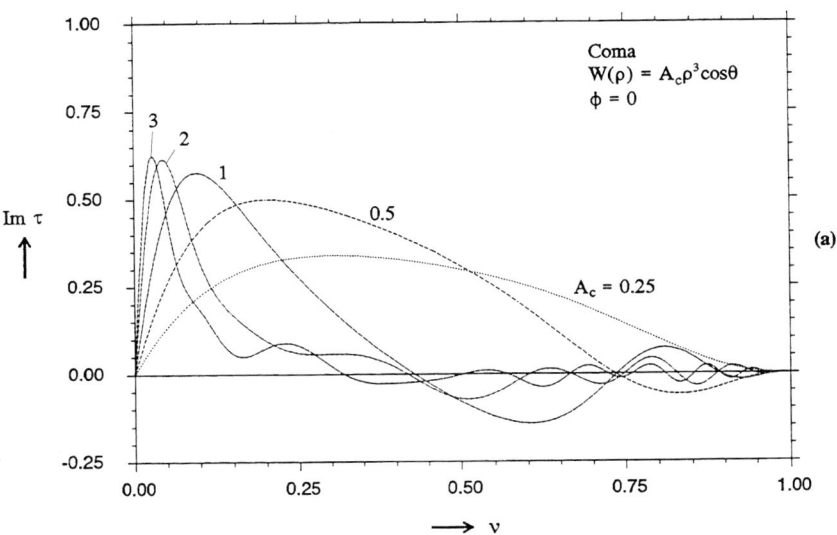

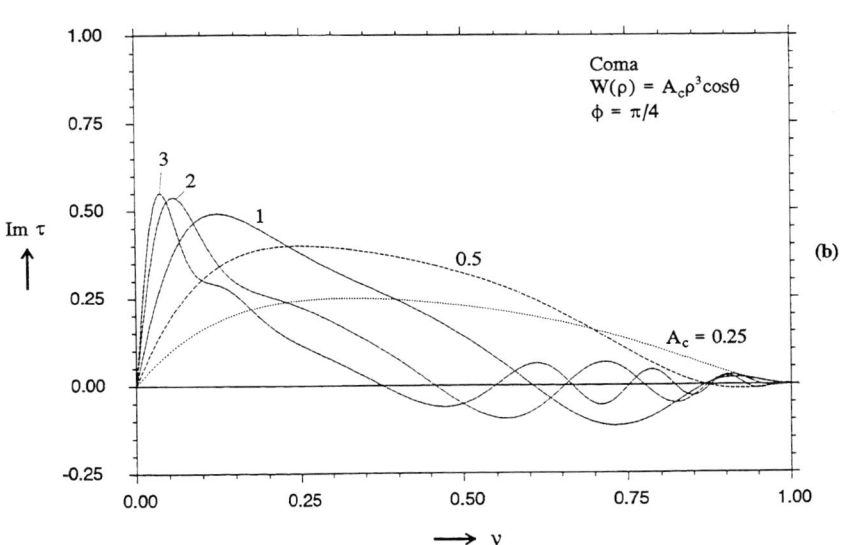

Figure 8–22. Imaginary part of the OTF of a system aberrated by coma. (a) $\phi = 0$, (b) $\phi = \pi/4$.

References

1. M. Born and E. Wolf, *Principles of Optics*, 5th ed. (Pergammon, New York, 1975); also J. W. Goodman, *Introduction to Fourier Optics*, (McGraw-Hill, New York, 1968).

2. G. B. Airy, "On the diffraction of an object-glass with circular aperture," Trans. Camb. Phil. Soc. **5**, 283-291 (1835).

3. V. N. Mahajan, "Axial irradiance and optimum focusing of laser beams," Appl. Opt. **22**, 3042-3053 (1983).

4. A. Maréchal, "Etude des effets combines de la diffraction et des aberrations geometriques sur l'image d'un point lumineux," Rev. d'Opt. **26**, 257-277 (1947).

5. B. R. A. Nijboer, "The diffraction theory of aberrations," Ph.D. thesis (University of Groningen, Groningen, The Netherlands, 1942), p. 17.

6. V. N. Mahajan, "Strehl ratio for primary aberrations in terms of their aberration variance," J. Opt. Soc. Am. **73**, 860-861 (1983).

7. V. N. Mahajan, "Strehl ratio for primary aberrations: some analytical results for circular and annular pupils," J. Opt. Soc. Am. **72**, 1258-1266 (1982).

8. V. N. Mahajan, "Line of sight of an aberrated optical system," J. Opt. Soc. Am. **A2**, 833-846 (1985).

9. W. B. King, "Dependence of the Strehl ratio on the magnitude of the variance of the wave aberration," J. Opt. Soc. Am. **58**, 655-661 (1968).

10. Lord Rayleigh, Phil. Mag. (5) **8**, 403 (1879); also his *Scientific Papers*, Vol. 1 (Dover, New York, 1964), p. 432.

11. V. N. Mahajan, "Aberrated point spread functions for rotationally symmetric aberrations," Appl. Opt. **22**, 3035-3041 (1983); also S. Szapiel, "Aberration-variance-based formula for calculating point-spread functions: rotationally symmetric aberrations," Appl. Opt. **25**, 244-251 (1986).

12. H. H. Hopkins, "The aberration permissible in optical systems," Proc. Phys. Soc. (London) **B52**, 449-470 (1957).

13. S. Szapiel, "Hopkins variance formula extended to low relative modulations," Optica Acta **33**, 981-999 (1986).

14. W. B. King, "Correlation between the relative modulation function and the magnitude of the wave aberration difference function," J. Opt. Soc. Am. **59**, 692-697 (1969).

CHAPTER 9

Optical Systems with Annular and Gaussian Pupils

9.1. Introduction

In Chapter 8 we have considered optical systems with circular exit pupils. Now we consider systems with *annular pupils*, for example, a Cassegrain telescope in which its secondary mirror *obscures* the central portion of its primary mirror. As in the case of a system with a circular pupil, we discuss the aberration-free PSF, axial irradiance, and the Strehl ratio of a system with an annular pupil. We show that the radius of the central bright spot of the PSF decreases, its principal or central maximum decreases in its value, and the secondary maxima increase in their values as the obscuration increases. However, the tolerance for a given Strehl ratio increases or decreases depending on the type of the aberration. The aberrated PSFs and OTFs are not discussed. Optical systems with circular pupils and *Gaussian illumination* across them are also considered along similar lines. For these systems, it is shown that the tolerance for an aberration increases compared to the corresponding tolerance for a system with a uniformly illuminated circular pupil. Finally, systems with *weakly truncated Gaussian pupils*, i.e., those having a very wide pupil compared to the width or the radius of the Gaussian illumination, are considered. In this case, the tolerance for a primary aberration is obtained in terms of its peak value at the *Gaussian radius* rather than at the edge of the pupil.

9.2. Annular Pupils

In this section, we discuss the imaging characteristics of systems with annular pupils. The aberration-free PSF, encircled power, axial irradiance, and Strehl ratio are discussed for increasing value of the obscuration of the pupil. The results obtained are compared with the corresponding results for systems with circular pupils.

9.2.1 Aberration-Free PSF

Consider a system with an annular exit pupil having inner and outer radii of ϵa and a, where ϵ is called its *obscuration ratio*. The PSF of the system, i.e., the irradiance distribution of the image of a point object formed by it, is given by Eq. (8-1) except that now the lower limit in the radial integration is ϵ instead of zero. The aberration-free PSF thus obtained is given by

$$I(r; \epsilon) = \frac{1}{(1 - \epsilon^2)^2} \left[\frac{2J_1(\pi r)}{\pi r} - \epsilon^2 \frac{2J_1(\pi \epsilon r)}{\pi \epsilon r} \right]^2, \qquad (9\text{-}1)$$

where $J_1(\bullet)$ is the first-order Bessel function of the first kind. It is normalized to unity at the center $r = 0$ by the central irradiance $PS_p/\lambda^2 R^2$, where P is the total power transmitted by the annular pupil and $S_p = \pi a^2(1 - \epsilon^2)$ is its clear area, λ is the wavelength of object radiation and R is the distance between the pupil plane and the image plane.

Note that r is in units of λF as in the case of a circular pupil, where $F = R/2a$ is the *focal ratio* or the *f-number* of the image-forming light cone. For a given total power P, the value of the *central maximum* decreases as $1 - \epsilon^2$ as ϵ increases due to the decrease in the clear pupil area. However, if the irradiance of the pupil is held constant, then the total power P also decreases as $1 - \epsilon^2$ and, therefore, the central irradiance decreases as $(1 - \epsilon^2)^2$ as ϵ increases.

The *minima* of the distribution have a value of zero at r values given by

$$J_1(\pi r) = \epsilon J_1(\pi \epsilon r), r \neq 0 . \qquad (9\text{-}2a)$$

Its *maxima* occur at r values given by

$$J_2(\pi r) = \epsilon^2 J_2(\pi \epsilon r), r \neq 0 , \qquad (9\text{-}2b)$$

where $J_2(\bullet)$ is a second-order Bessel function of the first kind. By integrating the irradiance distribution across a circle of radius r_c we obtain the encircled power $P(r_c)$. Both the irradiance and encircled-power distributions are shown in Figure 9-1 for several values of ϵ. We note that the radius of the central bright disc (first dark ring corresponding to the first minimum) decreases as ϵ increases. It can be shown that as $\epsilon \rightarrow 1$, the irradiance distribution approaches $J_0^2(\pi r)$. Its first zero occurs at 0.76 compared to a value of 1.22 [first zero of $J_1(\pi r)$] when $\epsilon = 0$. The values of the secondary maxima of a distribution relative to the value of its principal maximum at $r = 0$ become higher as ϵ increases. For example, when $\epsilon = 0.5$, the first secondary maximum has a value of 9.63% of the principal maximum compared to a value of 1.75% for a circular pupil. The minima, maxima, and the corresponding irradiances and encircled powers are given in Table 9-1 for $\epsilon = 0(0.1)0.9$.

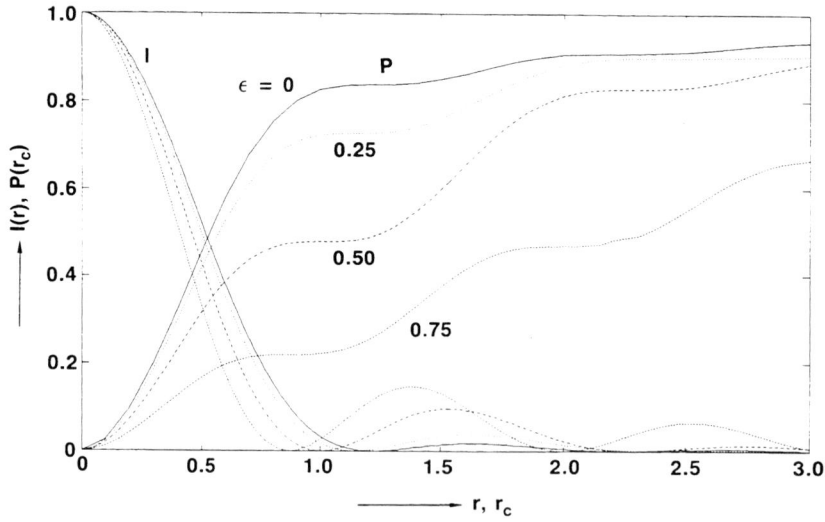

Figure 9–1. Irradiance and encircled-power distributions for an annular pupil. ϵ is the obscuration ratio of the pupil.

An interesting observation[1] comes about when the irradiance distribution is considered for large values of r and large values of ϵ. Figure 9-2 shows the distributions for $\epsilon = 0, 0.8$, and 1. A corresponding picture of these distributions is shown in Figure 9-3. We note that, for a circular pupil, the distribution consists of maxima and minima indicating a bright central disc surrounded by dark and bright rings. The successive maxima decrease in value monotonically. However, for an annular pupil, the distribution consists of not only the dark and bright rings but also of a *periodic ring group structure*. The number of maxima in a period is given by $n = 2/(1 - \epsilon)$, which is equal to the ratio of the outer diameter and the width of the annulus, provided that n is an integer. The distribution appears to be divided into ring groups. The group minima are the lowest ring maxima and correspond to ring numbers that are multiples of n, e.g., 10, 20, 30, etc., for $\epsilon = 0.8$. The radius of a ring group is also a multiple of n (in units of λF) since the spacing between two successive maxima or minima is approximately unity. The central bright spot or the first dark ring of radius 1.22 contains 83.8% of the total power in the image when $\epsilon = 0$. For $\epsilon = 0.8$, as may be seen from Table 9-1, the first dark ring has a radius of 0.85 and contains only 17.2% of the total power. However, the central ring group in this case has a radius of 10.10 and contains 90.3% of the total power. When n is not an integer, then the distribution becomes complex. For example, for $\epsilon = 0.7, n = 6.67$, and the distribution has a double periodicity with the number of maxima in the two periods equal to 6 and 7 (two integers closest to n).

9.2.2 Axial Irradiance

The axial irradiance of the image-forming beam for an aberration-free system with an annular pupil may be obtained in the same manner as for a system with a circular pupil. Thus, we let $r = 0$, $\Phi(\rho) = A_d \rho^2$, and replace the lower limit of radial integration from 0 to ϵ in Eq. (8-1), thereby obtaining the result

$$I(0; z; \epsilon) = \frac{PS_p}{\lambda^2 z^2} \left\{ \frac{\sin[A_d(1 - \epsilon^2)/2]}{A_d(1 - \epsilon^2)/2} \right\}^2 . \tag{9-3a}$$

Equation (9-3a) differs from the corresponding Eq.(8-7) for systems with circular pupils in that the quantity A_d in the latter has been replaced by $A_d(1 - \epsilon^2)$. It represents the peak defocus phase aberration at the outer edge of the annular pupil relative to its value at the inner edge. Accordingly, the defocus tolerance for a given Strehl ratio for a system with an annular pupil is *larger* by a factor of $(1 - \epsilon^2)^{-1}$ compared to its corresponding value if ϵ were zero. The axial irradiance is minimum and equal to zero at z values given by

$$R/z = 1 + [2n/N(1 - \epsilon^2)], \quad n = \pm 1, \pm 2, \ldots , \tag{9-3b}$$

where $N = a^2/\lambda R$ is the Fresnel number of the pupil if ϵ were zero. The maxima of axial irradiance, obtained by equating the derivative of Eq. (9-3a) with respect to z equal to zero, are given by the solutions of

$$\tan[A_d(1 - \epsilon^2)/2] = (R/z)A_d(1 - \epsilon^2)/2, \quad z \neq R . \tag{9-3c}$$

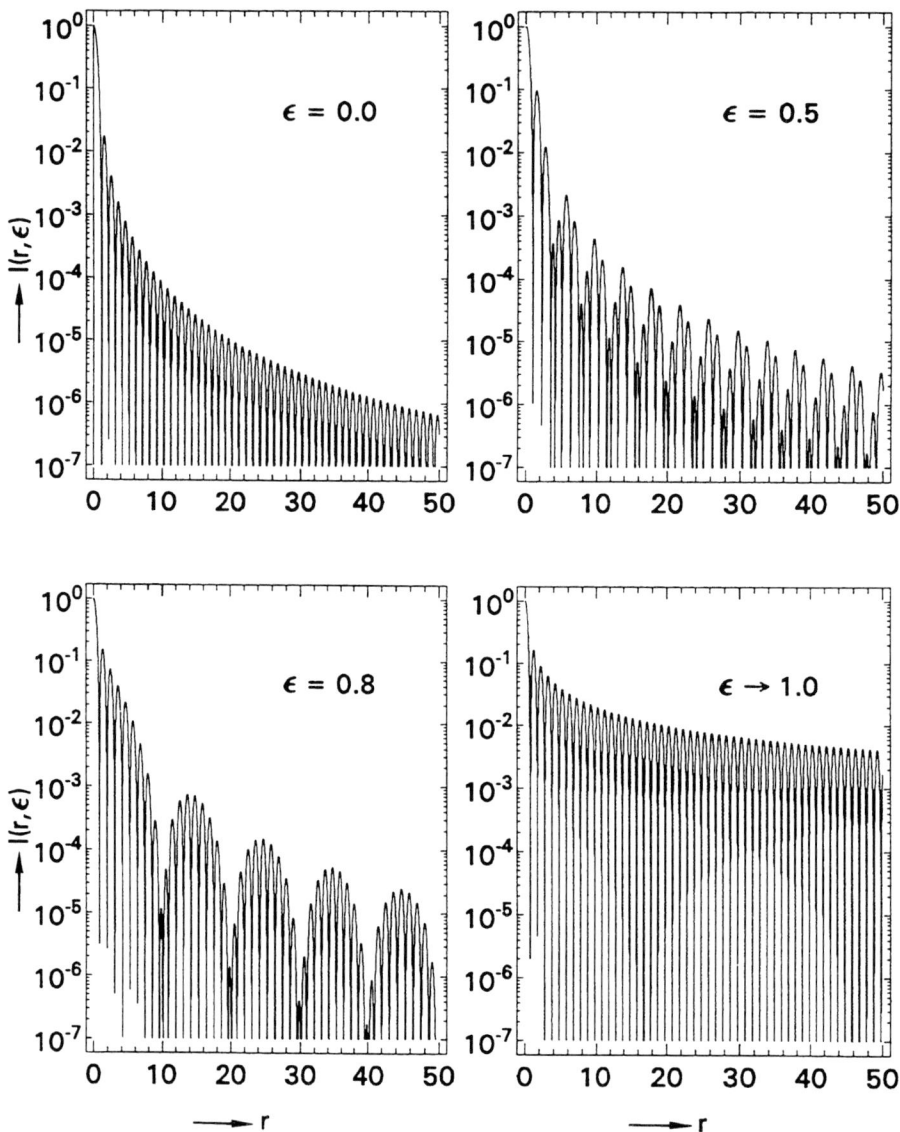

Figure 9–2. Irradiance distributions for circular ($\epsilon = 0$) and annular ($\epsilon \neq 0$) pupils. The case $\epsilon \to 1$ represents the limiting case of a totally obscured pupil.

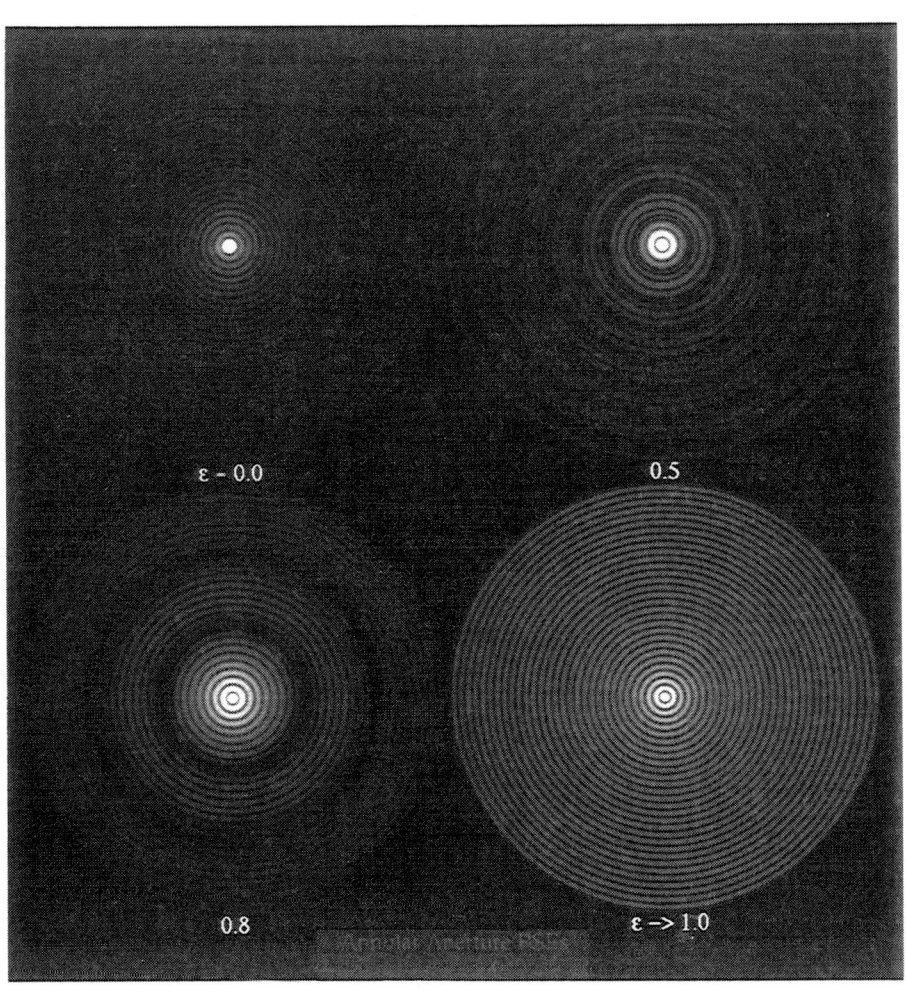

Figure 9–3. Pictures of irradiance distributions for circular and annular pupils. ϵ **is the obscuration ratio of the pupil.**

Table 9–1a. Positions r of PSF maxima and minima for an annular pupil in units of λF and corresponding irradiance and encircled power.

Max/Min	ε = 0			ε = 0.1			ε = 0.2			ε = 0.3			ε = 0.4		
	r, r_c	$I(r)$	$P(r_c)$	r, r_c	$I(r)$	$P(r_c)$	r, r_c	$I(r)$	$P(r_c)$	r, r_c	$I(r)$	$P(r_c)$	r, r_c	$I(r)$	$P(r_c)$
Max	0	1	0	0	1	0	0	1	0	0	1	0	0	1	0
Min	1.22	0	0.838	1.21	0	0.818	1.17	0	0.764	1.11	0	0.682	1.06	0	0.584
Max	1.63	0.0175	0.867	1.63	0.0206	0.853	1.63	0.0304	0.818	1.61	0.0475	0.766	1.58	0.0707	0.702
Min	2.23	0	0.910	2.27	0	0.906	2.36	0	0.900	2.42	0	0.899	2.39	0	0.885
Max	2.68	0.0042	0.922	2.68	0.0031	0.914	2.69	0.0015	0.904	2.73	0.0011	0.902	2.77	0.0033	0.893
Min	3.24	0	0.938	3.18	0	0.925	3.09	0	0.908	3.10	0	0.904	3.30	0	0.903
Max	3.70	0.0016	0.944	3.70	0.0024	0.936	3.68	0.0037	0.926	3.64	0.0028	0.916	3.66	0.0007	0.905
Min	4.24	0	0.952	4.32	0	0.949	4.37	0	0.947	4.22	0	0.929	4.04	0	0.907
Max	4.71	0.0008	0.957	4.71	0.0004	0.951	4.74	0.0004	0.949	4.75	0.0016	0.938	4.66	0.0028	0.922
Min	5.24	0	0.961	5.15	0	0.953	5.16	0	0.951	5.42	0	0.949	5.31	0	0.939
Max	5.72	0.0004	0.964	5.71	0.0008	0.959	5.69	0.0006	0.955	5.73	0.0001	0.950	5.79	0.0008	0.944
Min	6.24	0	0.968	6.35	0	0.965	6.23	0	0.959	6.07	0	0.950	6.43	0	0.950
Max	6.72	0.0003	0.970	6.73	0.0001	0.966	6.74	0.0004	0.962	6.67	0.0006	0.955	6.72	0.0001	0.950
Min	7.25	0	0.972	7.14	0	0.967	7.35	0	0.966	7.27	0	0.961	7.03	0	0.950
Max	7.73	0.0002	0.974	7.72	0.0003	0.970	7.72	0.0001	0.967	7.77	0.0003	0.963	7.65	0.0004	0.954
Min	8.25	0	0.975	8.34	0	0.974	8.11	0	0.967	8.38	0	0.966	8.22	0	0.958
Max	8.73	0.0001	0.977	8.74	0.0001	0.975	8.72	0.0003	0.971	8.72	0.0000	0.966	8.77	0.0004	0.962
Min	9.25	0	0.978	9.16	0	0.975	9.38	0	0.974	9.06	0	0.967	9.46	0	0.966
Max	9.73	0.0001	0.979	9.72	0.0001	0.977	9.75	0.0000	0.975	9.70	0.0002	0.970	9.78	0.0000	0.966
Min	10.25	0	0.980	10.30	0	0.979	10.16	0	0.975	10.32	0	0.973	10.13	0	0.966

Table 9–1b. Positions r of PSF maxima and minima for an annular pupil in units of λF and corresponding irradiance and encircled power.

Max/ Min	ε	0.5 r, r_c	0.5 $I(r)$	0.5 $P(r_c)$	0.6 r, r_c	0.6 $I(r)$	0.6 $P(r_c)$	0.7 r, r_c	0.7 $I(r)$	0.7 $P(r_c)$	0.8 r, r_c	0.8 $I(r)$	0.8 $P(r_c)$	0.9 r, r_c	0.9 $I(r)$	0.9 $P(r_c)$
Max		0	1	0	0	1	0	0	1	0	0	1	0	0	1	0
Min		1.000	0	0.479	0.95	0	0.372	0.90	0	0.269	0.85	0	0.172	0.81	0	0.082
Max		1.54	0.0963	0.618	1.48	0.1203	0.512	1.41	0.1395	0.389	1.35	0.1527	0.256	1.28	0.1600	0.124
Min		2.29	0	0.829	2.17	0	0.717	2.06	0	0.560	1.95	0	0.376	1.85	0	0.184
Max		2.76	0.0124	0.859	2.69	0.0306	0.784	2.58	0.0533	0.649	2.47	0.0734	0.456	2.35	0.0861	0.229
Min		3.49	0	0.901	3.39	0	0.873	3.22	0	0.761	3.05	0	0.554	2.90	0	0.284
Max		3.78	0.0004	0.902	3.84	0.0045	0.886	3.74	0.0192	0.808	3.57	0.0401	0.619	3.40	0.0566	0.328
Min		4.12	0	0.903	4.52	0	0.902	4.38	0	0.865	4.16	0	0.695	3.95	0	0.379
Max		4.60	0.0009	0.907	4.80	0.0001	0.903	4.86	0.0050	0.880	4.68	0.0218	0.741	4.46	0.0404	0.421
Min		5.05	0	0.910	5.11	0	0.903	5.52	0	0.899	5.27	0	0.795	5.00	0	0.468
Max		5.66	0.0022	0.923	5.58	0.0004	0.905	5.91	0.0005	0.901	5.78	0.0110	0.824	5.51	0.0299	0.507
Min		6.30	0	0.938	6.00	0	0.906	6.47	0	0.903	6.37	0	0.857	6.05	0	0.549
Max		6.81	0.0008	0.943	6.61	0.0016	0.916	6.72	0.0000	0.903	6.87	0.0048	0.872	6.56	0.0224	0.584
Min		7.50	0	0.950	7.19	0	0.925	6.97	0	0.903	7.47	0	0.889	7.10	0	0.622
Max		7.79	0.0000	0.950	7.75	0.0013	0.934	7.53	0.0004	0.905	7.95	0.0016	0.894	6.61	0.0169	0.652
Min		8.12	0	0.950	8.40	0	0.944	7.98	0	0.906	8.57	0	0.901	8.16	0	0.685
Max		8.62	0.0001	0.951	8.87	0.0004	0.947	8.58	0.0010	0.913	8.98	0.0003	0.902	8.67	0.0127	0.711
Min		9.05	0	0.952	9.53	0	0.950	9.13	0	0.919	9.58	0	0.903	9.21	0	0.739
Max		9.68	0.0004	0.957	9.80	0.0000	0.950	9.69	0.0011	0.927	9.83	0.0000	0.903	9.72	0.0094	0.761
Min		10.31	0	0.962	10.11	0	0.950	10.28	0	0.935	10.10	0	0.903	10.26	0	0.784

Figure 9-4 shows how the axial irradiance of an annular beam with $\epsilon = 0.5$ varies for $N = 1$, 10, and 100. Comparing it with Figure 8-2, we note that the effect of the obscuration is to reduce the irradiance at the principal maximum but to increase it at the secondary maxima. Also, the maxima and minima occur at smaller z values for an annular pupil. As in the case of circular beams, the axial irradiance of annular beams also becomes symmetric about the focal point $z = R$ as N increases.

9.2.3 Strehl Ratio

For small aberrations, the Strehl ratio of an aberrated image is still given by Eqs. (8-13)–(8-15), except that the variance σ_Φ^2 of the aberration $\Phi(\rho,\theta;\epsilon)$ is across the annular region of the pupil. This in turn implies that the mean and the mean square values of the aberration are given by

$$< \Phi^n > \; = \; \frac{1}{\pi(1-\epsilon^2)} \int_\epsilon^1 \int_0^{2\pi} \Phi^n(\rho,\theta;\epsilon)\rho d\rho d\theta, \qquad (9\text{-}4)$$

with $n = 1$ and 2, respectively.

The form of a primary aberration and its corresponding standard deviation are listed in Table 9-2. The balanced aberrations listed in the table represent balancing of an aberration with another to minimize its variance across the annular pupil. They can be identified with the corresponding *Zernike annular polynomials*.[2]

Figure 9-5 shows how the standard deviation of an aberration, for a given value of the aberration coefficient A_i, varies with the obscuration ratio of the pupil. In Figures 9-5a and 9-5b, the amounts of defocus and tilt required to minimize the variance of spherical aberration and coma, respectively, are also shown. We observe from these figures that the standard deviation of spherical and balanced spherical aberrations and defocus decreases as ϵ increases. Correspondingly, the tolerance in terms of their aberration coefficients A_s and A_d, for a given Strehl ratio, increases. The standard deviation of coma, astigmatism, balanced astigmatism, and tilt increases as ϵ increases. The standard deviation of balanced coma first slightly increases, achieves its maximum value at $\epsilon = 0.29$, and then decreases rapidly as ϵ increases. The factor by which the standard deviation of an aberration is reduced by balancing it with another aberration decreases in the case of spherical aberration and coma, but increases in the case of astigmatism, as ϵ increases.

Figures 9-6 show how the Strehl ratio of a primary aberration varies with its standard deviation for $\epsilon = 0.5$ and 0.75. Approximate as well as exact results are shown in these figures.[3] The exact results are obtained by the use of Eq. (8-8) except that the lower limit in the radial integral is ϵ and not zero.[4] The curves for a given aberration and for the corresponding balanced aberration can be distinguished from each other by their behavior for large σ_w values (near 0.25λ). For example, coma is shown by the evenly dashed curves; the higher dashed curve is for coma and the lower is for balanced coma.

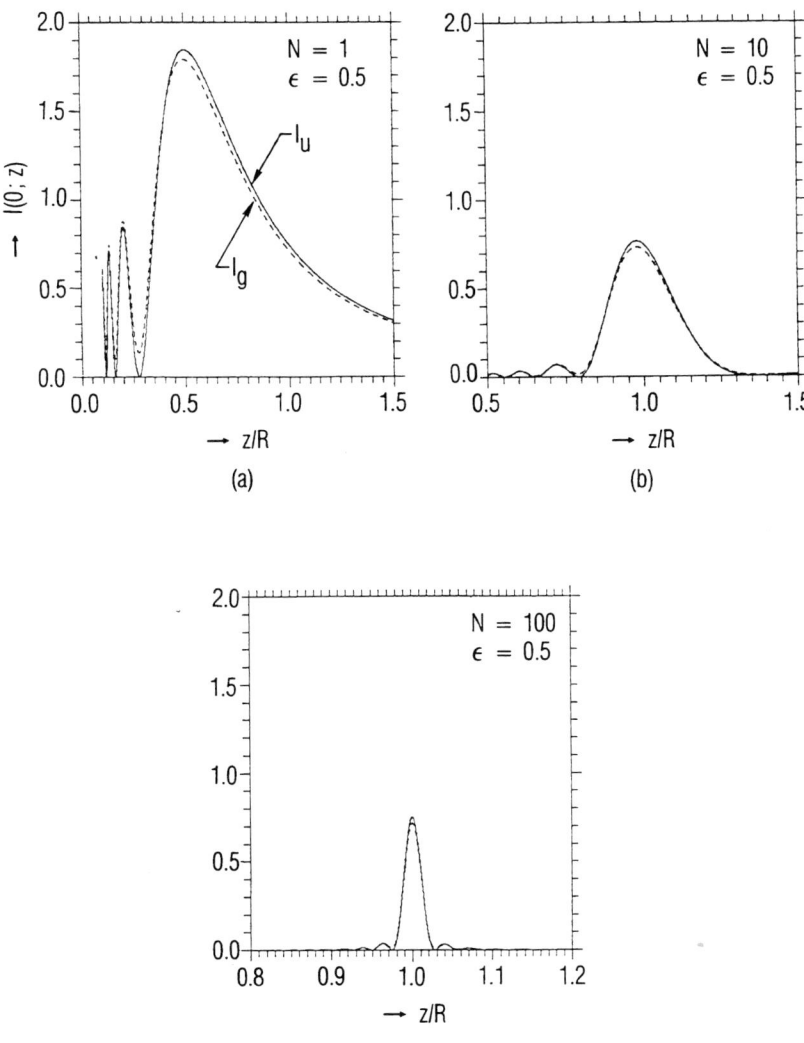

Figure 9–4. Axial irradiance of an annular beam focused at a distance R with a Fresnel number $N = a^2/\lambda R$. The subscripts u and g refer to uniform and Gaussian beams, respectively (from Reference 4).

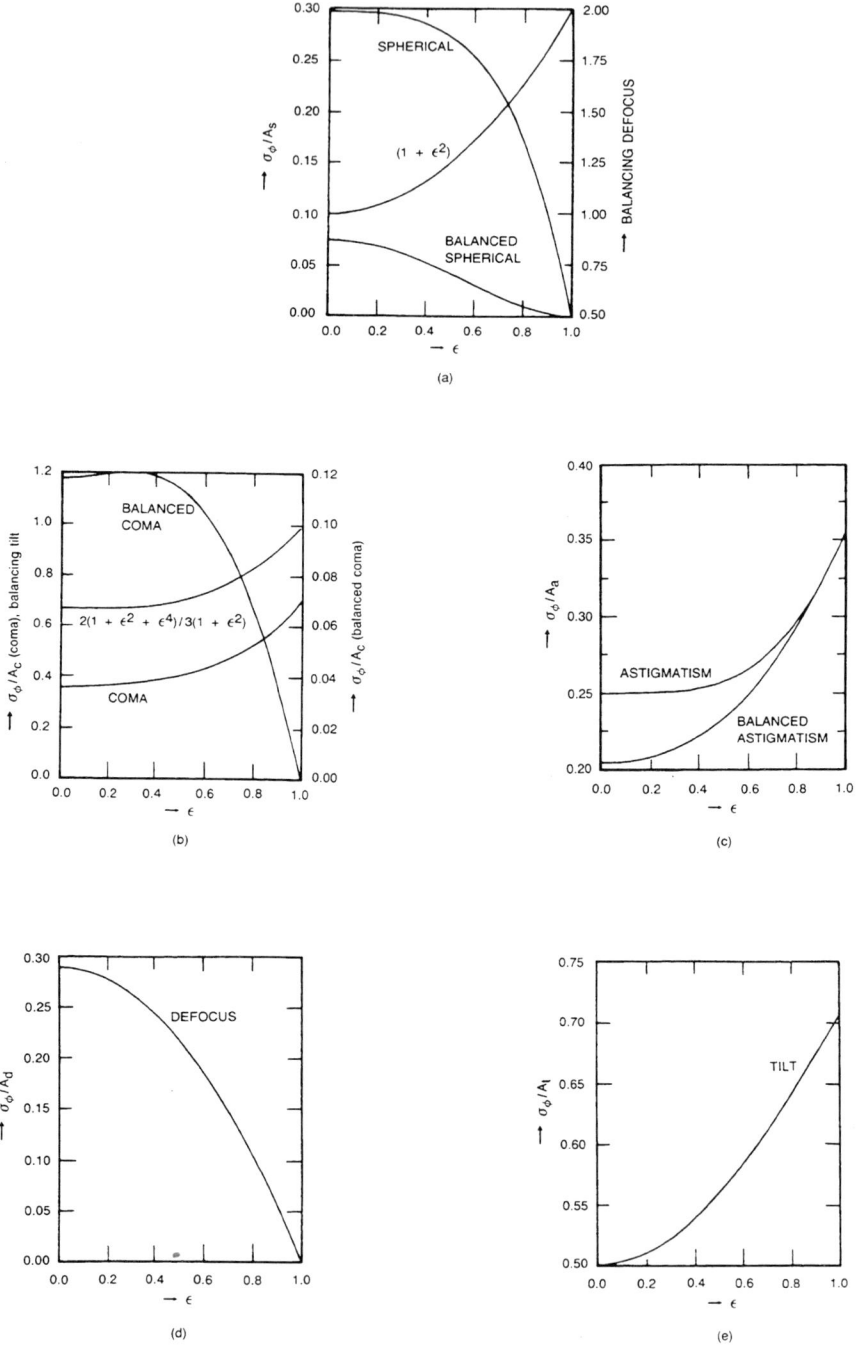

Figure 9–5. Variation of standard deviation of a primary and balanced primary aberration with obscuration ratio ε. Variation of balancing defocus in the case of spherical aberration and tilt in the case of coma are also shown. (a) Spherical aberration, (b) coma, (c) astigmatism, (d) defocus, and (e) tilt (from Reference 2).

Table 9–2. Primary aberrations and their standard deviations for optical systems with annular pupils.

Aberration	$\Phi(\rho,\theta;\epsilon)$	σ_Φ
Spherical	$A_s\rho^4$	$\dfrac{1}{3\sqrt{5}}(4-\epsilon^2-6\epsilon^4-\epsilon^6+4\epsilon^8)^{1/2}A_s$
Balanced Spherical	$A_s[\rho^4-(1+\epsilon^2)\rho^2]$	$\dfrac{1}{6\sqrt{5}}(1-\epsilon^2)^2A_s$
Coma	$A_c\rho^3\cos\theta$	$\dfrac{1}{\sqrt{8}}(1+\epsilon^2+\epsilon^4+\epsilon^6)^{1/2}A_c$
Balanced Coma	$A_c\left(\rho^3-\dfrac{2}{3}\dfrac{1+\epsilon^2+\epsilon^4}{1+\epsilon^2}\rho\right)\cos\theta$	$\dfrac{(1-\epsilon^2)(1+4\epsilon^2+\epsilon^4)^{1/2}}{6\sqrt{2}\,(1+\epsilon^2)^{1/2}}A_c$
Astigmatism	$A_a\rho^2\cos^2\theta$	$\dfrac{1}{4}(1+\epsilon^4)^{1/2}A_a$
Balanced Astigmatism	$A_a\rho^2(\cos^2\theta-1/2)$	$\dfrac{1}{2\sqrt{6}}(1+\epsilon^2+\epsilon^4)^{1/2}A_a$
Defocus	$A_d\rho^2$	$\dfrac{1}{2\sqrt{3}}(1-\epsilon^2)A_d$
Tilt	$A_t\rho\cos\theta$	$\dfrac{1}{2}(1+\epsilon^2)^{1/2}A_t$

As in the case of circular pupils, the expressions for S_1 and S_2 underestimate the true Strehl ratio. The expression for S_3 overestimates the true Strehl ratio for $\epsilon \geq 0.5$. It gives the Strehl ratio with an error of less than 10% for $S \geq 0.4$. For smaller obscurations, the error is less than 10% for $S \geq 0.3$. The percent error is defined as $100(1 - S_3/S)$.

Using S_1 to estimate the Strehl ratio, Figure 9-7 shows how the aberration coefficient A_i of a primary aberration for 10% error varies with the obscuration ratio.[5] It is evident that this coefficient increases with obscuration in the case of spherical, balanced spherical, and balanced coma, but decreases in the case of astigmatism, balanced astigmatism, and coma. When the aberration coefficient A_i of an aberration is equal to a quarter wave, the variation of the corresponding Strehl ratio with ϵ is shown in Figure 9-8. It is evident that a Strehl ratio of 0.8 is obtained in very few cases. Comparing this figure with Figures 9-6, we again conclude, as in the case of circular pupils, that it is advantageous to use the standard deviation of an aberration instead of the aberration coefficient to estimate the Strehl ratio. For example, a Strehl ratio of 0.8 is obtained for any aberration with a standard deviation of $\sigma_w = \lambda/14$. On the other hand this value of Strehl ratio is obtained for different values of the aberration coefficient for different aberrations.

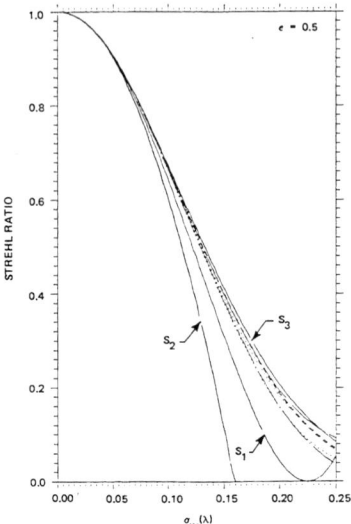

Figure 9–6a. Strehl ratio for annular pupils with ε = 0.5 as a function of the standard deviation σ_w of an aberration in units of λ. The Strehl ratio in the case of coma is practically independent of whether or not it is balanced. For large values of σ_w, the Strehl ratio for astigmatism is larger than that for balanced astigmatism. Spherical....., Coma———, Astigmatism –·– (from Reference 2).

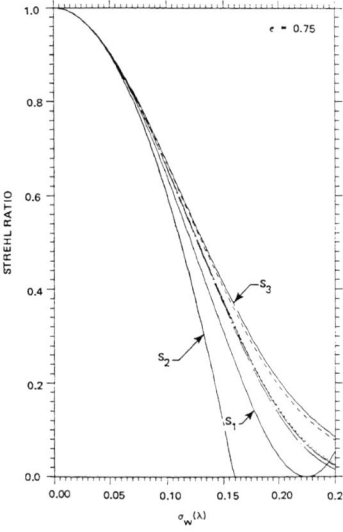

Figure 9–6b. Strehl ratio for annular pupils with ε = 0.75 as a function of the standard deviation σ_w in units of λ. For large values of σ_w, the Strehl ratio for balanced coma is higher than that for coma. The opposite is true for astigmatism. Note that the curves for coma and astigmatism are practically identical. Spherical....., Coma———, Astigmatism –·– (from Reference 2).

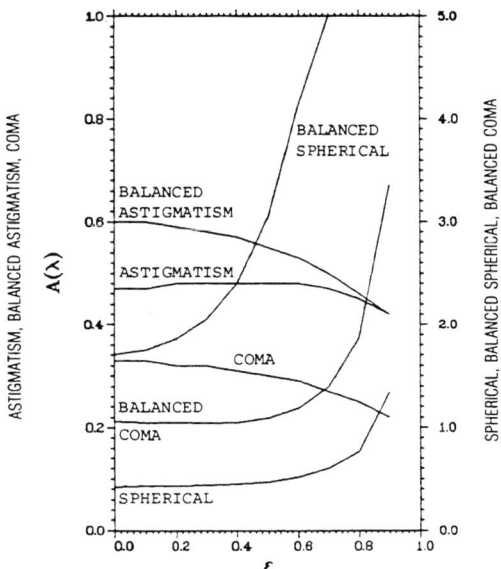

Figure 9–7. Variation of a primary aberration coefficient A_i with obscuration ratio ϵ for 10% error when S_1 is used to estimate the Strehl ratio (from Reference 3).

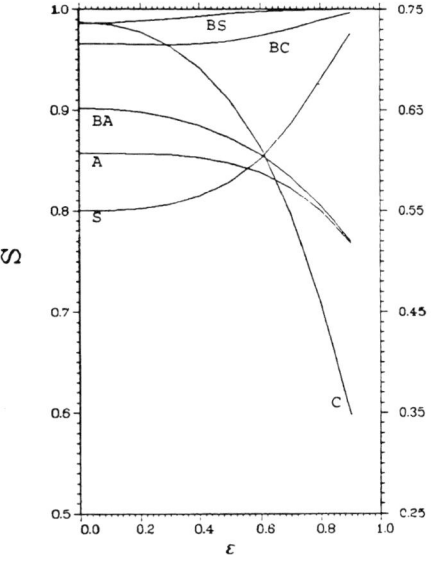

Figure 9–8. Strehl ratio for $A_i = \lambda/4$ as a function of ϵ. S, spherical; BS, balanced spherical; C, coma, BC, balanced coma; A, astigmatism, and BA, balanced astigmatism. The right-hand side vertical scale is only for coma (from Reference 3).

9.3. Gaussian Pupils

So far we have considered optical systems which have *uniform* amplitude across their exit pupils. Now we consider systems with exit pupils having nonuniform amplitude across them in the form of a *Gaussian*.[2,5] Such pupils are often referred to as *Gaussian pupils*. The Gaussian amplitude may, for example, be obtained by placing a filter with Gaussian transmission at the pupil. A system with a nonuniform amplitude across its pupil is called an *apodized system*. The motivation for apodizing a system is to reduce the values of the secondary maxima of its PSF relative to the value of the principal maximum. The discussion given here applies equally well to the propagation of *Gaussian laser beams*. For a Gaussian pupil transmitting the same total power as a circular pupil with uniform transmission, the central value of the PSF is smaller and the tolerance for an aberration is higher.

9.3.1. Aberration–Free PSF

The Gaussian amplitude may be written

$$A(\rho) = A_0 \exp(-\gamma \rho^2) , \tag{9-5}$$

where A_0 is a constant and γ is a parameter that defines truncation of the Gaussian by the pupil. If we let ω be the radial distance at which the amplitude drops to $1/e$ of its value at the center, then $\gamma = (a/\omega)^2$, where a is the radius of the exit pupil. We will call ω the *Gaussian radius*. In the limit $\gamma \to 0$, we obtain a uniformly illuminated pupil. The total power transmitted by the pupil is obtained by integrating $A^2(\rho)$ across the pupil.

The PSF for a Gaussian pupil or the irradiance distribution of a focused Gaussian beam may be obtained from Eq. (8-1) provided the Gaussian amplitude is inserted under the integral in this equation. As an example, Figure 9-9 shows the irradiance and encircled–power distributions for $\gamma = 1$. For comparison, the corresponding distributions for a uniform beam having the same total power as the Gaussian are also shown. The subscripts u and g refer to uniform and Gaussian beams, respectively. At and near the focal point, a uniform beam gives a higher irradiance than a Gaussian beam. Thus, $I_u > I_g$ for $r < 0.42$. For larger values of r, $I_g > I_u$, except in the secondary rings, where again $I_u > I_g$. The encircled power $P_u \gtrsim P_g$ for $r_c \gtrsim 0.63$. Of course, as $r_c \to \infty$, $P_u \to P_g \to 1$. The Gaussian illumination broadens the central disc but reduces the power in the secondary rings.

For clarity, the irradiance distributions are also plotted on a logarithmic scale. The positions of maxima and minima and the corresponding irradiance and encircled–power values are given in Table 9–3. Comparing them with those in Table 8–1, it is evident that the corresponding maxima and minima for a Gaussian beam are located at higher values of r than those for a uniform beam. Moreover, whereas the principal maximum for a Gaussian beam is only slightly lower (0.924 compared with 1), the secondary maxima are lower by a factor > 3 compared with the corresponding maxima for a uniform beam.

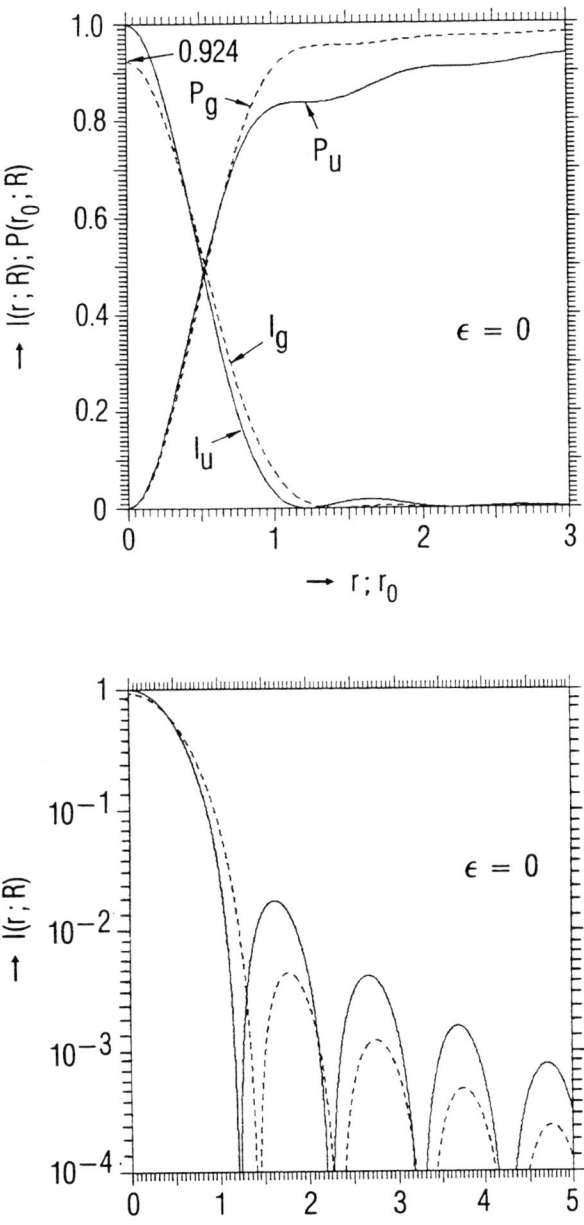

Figure 9–9. Focal-plane irradiance and encircled-power distributions for uniform and Gaussian ($\gamma = 1$) circular beams of a given total power P. The irradiance is in units of $PS_p/\lambda^2 R^2$ and the encircled power is in units of P. r and r_c are in units of λF (from Reference 4).

Table 9–3. Positions of maxima and minima of focal–plane irradiance distribution and corresponding encircled powers for a Gaussian circular beam with $\gamma = 1$ (from Reference 4).

Max/Min	r,r_c	$I(r)$	$P(r_c)$
Max	0	0.924	0
Min	1.43	0	0.955
Max	1.79	0.0044	0.962
Min	2.33	0	0.973
Max	2.76	0.0012	0.976
Min	3.30	0	0.981
Max	3.76	0.0005	0.983
Min	4.29	0	0.985
Max	4.75	0.0002	0.986

9.3.2 Axial Irradiance

Figure 9-10 shows how the axial irradiance of a focused Gaussian beam with $\gamma = 1$ differs from that of a focused uniform beam when the Fresnel number $N = 1, 10,$ and 100. We note that the principal maximum is higher for the uniform beam compared with that for the Gaussian beam. However, the secondary maxima are higher for the Gaussian beam. Moreover, whereas the axial minima for the uniform beam have a value of zero, the minima for the Gaussian beam have non-zero values. We note that the curves become symmetric about the focal point $z = R$ as N increases. It should be noted that even though the principal maximum of axial irradiance does not lie at the focus, unless N is very large, maximum central irradiance on a target at a given distance from the pupil is obtained when the beam is focused on it. (Similarly, for a weakly truncated Gaussian pupil discussed later, minimum Gaussian radius is obtained on a target when the beam is focused on it, even though a smaller radius occurs at a distance $z < R$ when N is small.[5])

9.3.3 Strehl Ratio

The Strehl ratio of an aberrated image is given by the approximate Eqs. (8-13)–(8-15), where the variance of the aberration is now across the amplitude-weighted pupil. Thus, for a circular pupil, the mean and the mean square values of the aberration are given by

$$< \Phi^n > \ = \ \int_0^1 \int_0^{2\pi} \exp(-\gamma \rho^2) \Phi^n(\rho,\theta) \rho \, d\rho \, d\theta \Big/ \int_0^1 \int_0^{2\pi} \exp(-\gamma \rho^2) \, \rho \, d\rho \, d\theta \ , \qquad (9\text{-}6)$$

with $n = 1$ and 2, respectively. Following the same procedure as for a uniformly illuminated circular pupil, we can obtain the balanced primary aberrations and their standard deviations. Table 9-4 gives the aberrations and their standard deviations for $\gamma = 1$, i.e., when $a = \omega$. Comparing these results with those given in Table 9-2 for $\epsilon = 0$, it is evident that the standard deviation of an aberration for a Gaussian pupil is somewhat smaller than the corresponding value for a uniform pupil. Accordingly, for a given small amount of aberration A_i, the Strehl ratio for a Gaussian pupil is somewhat higher than that for a uniform pupil. Similarly, for a given Strehl ratio, the aberration tolerance for a Gaussian pupil is somewhat higher than that for a uniform pupil. Moreover, the balancing defocus in the case of spherical aberration and the balancing tilt in the case of coma are somewhat smaller for a Gaussian pupil, compared to their corresponding values for a uniform pupil; i.e., the *diffraction focus* for these aberrations in the case of a Gaussian pupil is slightly different from the corresponding focus for a uniform pupil. We also note that, although aberration balancing in the case of a uniform pupil reduces the standard deviation of spherical aberration and coma by factors of 4 and 3, respectively, the reduction in the case of astigmatism is only a factor of 1.22. For a Gaussian pupil, the trend is similar but the reduction factors are smaller for spherical aberration and coma, and larger for astigmatism. They are 3.74, 2.64, and 1.27, corresponding to spherical aberration, coma, and astigmatism,* respectively.

Table 9–4. Primary aberrations and their standard deviations for optical systems with Gaussian circular pupils and $\gamma = 1$.

Aberration	$\Phi(\rho,\theta)$	σ_Φ
Spherical	$A_s\rho^4$	$A_s/3.67$
Balanced Spherical	$A_s(\rho^4 - 0.933\rho^2)$	$A_s/13.71$
Coma	$A_c\rho^3\cos\theta$	$A_c/3.33$
Balanced Coma	$A_c(\rho^3 - 0.608\rho)\cos\theta$	$A_c/8.80$
Astigmatism	$A_a\rho^2\cos^2\theta$	$A_a/4.40$
Balanced Astigmatism	$A_a\rho^2(\cos^2\theta - 1/2)$	$A_a/5.61$
Defocus	$A_d\rho^2$	$A_d/3.55$
Tilt	$A_t\rho\cos\theta$	$A_t/2.19$

9.3.4 Weakly Truncated Pupils

For a weakly truncated Gaussian pupil, i.e., for large values of γ, the upper limit on the radial variable in Eq. (9-5) and any associated equations may be changed from 1 to ∞ with negligible error. Numerical calculations show that for $\gamma \geq 9$ (or $a \geq 3\omega$), the difference between the exact PSF and the approximate result thus obtained may

---

*The factor for astigmatism is incorrectly stated as 1.16 in the text and 1.66 in Table 5 of Reference 5.

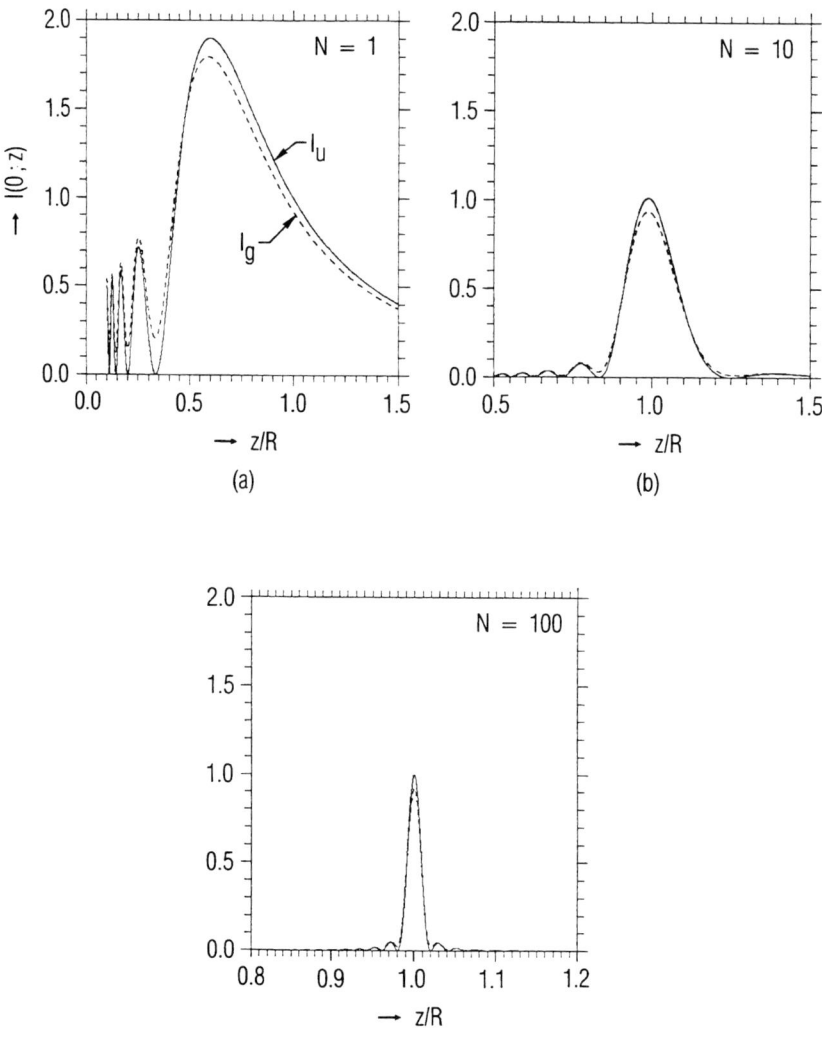

Figure 9–10. Axial irradiance of uniform and Gaussian ($\gamma = 1$) circular beams focused at a distance R with a Fresnel number $N = a^2/\lambda R = 1, 10, 100$. The irradiance is in units of the focal-point irradiance for a uniform circular beam (from Reference 4). The subscripts u and g refer to uniform and circular beams, respectively. This figure is the same as Figure 8-2.

be neglected.[5] Moreover, in the limit of an untrucated beam, an aberration-free Gaussian beam propagates as a Gaussian. The beam radius and the irradiance distribution in a plane at a distance z from a plane where its beam radius is ω are given by

$$\omega_z^2 = (\lambda z/\pi\omega)^2 + \omega^2(1 - z/R)^2 \tag{9-7}$$

and

$$I(r; z) = (2P/\pi\omega_z^2) \exp(-2r^2/\omega_z^2) , \tag{9-8}$$

respectively. In Eq. (9-8), r is the radial distance of a point in the observation plane from the axis of the beam without any normalization. For a weakly truncated beam, since the power in the pupil is concentrated in a small region near its center, the effect of the aberration in its outer region is negligible. Accordingly, the aberration tolerance in terms of the peak value of a primary aberration at the edge ($\rho = 1$) of the pupil is not very meaningful. It is more appropriate, for example, to consider the tolerance in terms of the peak value at the Gaussian radius. If we define

$$\rho' = \sqrt{\gamma}\,\rho , \tag{9-9}$$

then, $\rho' = 1$ corresponds to the Gaussian radius. Correspondingly, we define the aberration coefficients

$$A_s' = A_s/\gamma^2, \ A_c' = A_c/\gamma^{3/2}, \ A_a' = A_a/\gamma, \ A_d' = A_d/\gamma, \ A_t' = A_t/\sqrt{\gamma} , \tag{9-10}$$

which represent the peak values of aberrations at the Gaussian radius.

Table 9-5 lists the aberrations in terms of the radial variable ρ' and the aberration coefficients A_i'. The standard deviations of these aberrations across the Gaussian pupil are also given in this table. We note that the balancing of an aberration reduces

Table 9–5. Primary aberrations and their standard deviations for optical systems with weakly ($\gamma \geq 9$) truncated Gaussian circular pupils.

Aberration	$\Phi(\rho',\theta)$	σ_Φ	A_i' for $S = 0.8$
Spherical	$A_s'\,\rho'^4$	$2\sqrt{5}A_s'$	$\lambda/63$
Balanced Spherical	$A_s'(\rho'^4 - 4\rho'^2)$	$2A_s'$	$\lambda/28$
Coma	$A_c'\,\rho'^3\cos\theta$	$\sqrt{3}A_c'$	$\lambda/24$
Balanced Coma	$A_c'(\rho'^3 - 2\rho')\cos\theta$	A_c'	$\lambda/14$
Astigmatism	$A_a'\,\rho'^2\cos^2\theta$	$A_a'/\sqrt{2}$	$\lambda/10$
Balanced Astigmatism	$A_a'\,\rho'^2(\cos^2\theta - 1/2)$	$A_a'/2$	$\lambda/7$
Field Curvature (defocus)	$A_d'\,\rho'^2$	$\sqrt{3}A_d'$	$\lambda/24$
Distortion (tilt)	$A_t'\,\rho'\cos\theta$	$\sqrt{2}A_t'$	$\lambda/20$

its standard deviation by a factor of $\sqrt{5}, \sqrt{3}$, and $\sqrt{2}$ in the case of spherical aberration, coma, and astigmatism, respectively. The amount of a balancing aberration decreases as γ increases in the case of spherical aberration and coma but does not change in the case of astigmatism. For example, in the case of spherical aberration, the amount of balancing defocus for a weakly truncated Gaussian pupil is $(4/\gamma)$ times the corresponding amount for a uniform pupil. Similarly, in the case of coma, the balancing tilt for a weakly truncated Gaussian pupil is $(3/\gamma)$ times the corresponding amount for a uniform pupil. Aberration tolerances in terms of the aberration coefficients A_i' for a Strehl ratio of 0.8 are given in Table 9-5. The tolerances in terms of the coefficients A_i may be obtained by use of Eq. (9-10).

The procedure outlined here for determining aberration tolerances can be extended to systems with annular Gaussian pupils, and orthogonal polynomials, similar to Zernike polynomials, representing balanced aberrations can be obtained.[2,5]

References

1. H. F. Tschunko, "Imaging performance of annular apertures," Appl. Opt. **18**, 1820-1823 (1974).

2. V. N. Mahajan, "Zernike annular polynomials for imaging systems with annular pupils," J. Opt. Soc. Am. **71**, 75–85, 1408 (1981), and **A1**, 685 (1984).

3. V. N. Mahajan, "Strehl ratio for primary aberrations in terms of their aberration variance," J. Opt. Soc. Am. **73**, 860–861 (1983).

4. V. N. Mahajan, "Strehl ratio for primary aberrations: some analytical results for circular and annular pupils," J. Opt. Soc. Am. **72**, 1258–1266 (1982).

5. V. N. Mahajan, "Uniform versus Gaussian beams: a comparison of the effects of diffraction, obscuration, and aberrations," J. Opt. Soc. Am. **A3**, 470–485 (1986).

CHAPTER 10

Line of Sight of an Aberrated Optical System

10.1 Introduction

In this chapter we consider the *line of sight* (LOS) of an aberrated optical system. The LOS is assumed here to coincide with the *centroid* of its diffraction point-spread function (PSF). For an aberration-free system, it coincides with the center of the PSF. For an aberrated system, it depends on the various orders of its coma aberrations. Thus, a coma aberration not only reduces the central value of the PSF like any other aberration, but it also shifts its centroid. We consider here PSFs aberrated by primary coma and give numerical results on the location of their peaks and centroids.

10.2 Theory

The LOS of an aberration-free optical system coincides with the center of its diffraction PSF. For an aberrated system, let us define its LOS as the centroid of its aberrated PSF. Thus, if $I(x,y)$ represents the irradiance distribution of the aberrated image of a point object, its centroid representing the LOS error of the system is given by

$$<x> = \frac{1}{P} \int xI(x,y)dxdy \qquad (10\text{-}1a)$$

and

$$<y> = \frac{1}{P} \int xI(x,y)dxdy , \qquad (10\text{-}1b)$$

where P is the total power in the image. It can be shown that the centroid thus obtained is identical to that obtained from the *geometrical* PSF.[1]

For a system with a *circular* exit pupil, let its aberration function in terms of Zernike circle polynomials (see Section 8.3.7) be given by

$$W(\rho,\theta) = \sum_{n=0}^{\infty} \sum_{m=0}^{n} \epsilon_m \sqrt{2(n + 1)} R_n^m(\rho)(c_{nm} \cos m\theta + s_{nm} \sin m\theta) , \qquad (10\text{-}2)$$

where c_{nm} and s_{nm} are the Zernike aberration coefficients representing the standard deviations of the corresponding aberration terms across the pupil (with the exception of the piston term $n = 0 = m$, which has a standard deviation of zero). It can be shown that the centroid of its aberrated PSF for a uniformly illuminated pupil is given by[1]

$$<x> = 2F \sum_{n=1}^{\infty}{}' \sqrt{2(n + 1)} c_{n1} \qquad (10\text{-}3a)$$

131

and

$$<y> = 2F \sum_{n=1}^{\infty}{}' \sqrt{2(n + 1)}s_{n1} , \qquad (10\text{-}3b)$$

where F is the focal ratio or the f–number of the image–forming light cone and a prime indicates a summation over odd integral values of n. We note that only those aberrations contribute to the LOS *errors* that vary with θ as $\cos\theta$ and $\sin\theta$. Aberrations varying as $\cos\theta$ contribute to $<x>$, and those varying as $\sin\theta$ contribute to $<y>$. For a given value of c_{n1} or s_{n1}, an aberration of a higher order n gives a larger LOS error. Thus, two Zernike aberrations with $m = 1$ but different values of n having the same standard deviation give different LOS errors, even though they give (approximately) the same Strehl ratio. (See Section 8.3.1 for a relationship between the Strehl ratio and standard deviation of the aberration.)

If we consider an aberration in the form

$$W(\rho,\theta) = A_k \rho^k \cos\theta , \qquad (10\text{-}4)$$

where k is an odd integer, we find that

$$<x> = 2FA_k \qquad (10\text{-}5a)$$

$$<y> = 0 . \qquad (10\text{-}5b)$$

Thus, the LOS error depends on the value of the peak aberration A_k but not on k. We note that for $k = 1$, the aberration is a tilt and for $k = 3$ it is coma, but they both give the same LOS error if $A_1 = A_3$, even though the corresponding PSFs are completely different. The reason for the same LOS error is that for a uniform circular pupil, the centroid depends only on the aberration along the perimeter of the pupil[1] which depends on A_k, but not on k.

10.3 Numerical Results

Figure 10-1 shows the central profiles of the PSFs aberrated by coma varying from $A_3 = 0$ to $A_3 = 2\lambda$, normalized by the aberration-free central irradiance. Note that λ is the optical wavelength. The locations of the peak x_p and centroid $<x>$ of the aberrated PSFs are given in Table 10-1. The irradiances I_p and I_c at these points and $I(0,0)$ at the PSF center are also given in this table. For example, when $A_3 = 0.5\lambda$, the Strehl ratio of the PSF is approximately equal to 0.32, but its peak value of 0.87 lies at the point $(0.66,0)$ compared to a value of unity at the center $(0,0)$ of the corresponding aberration-free PSF. The centroid of the PSF lies at $(1,0)$. Thus, the centroid of the PSF shifts by an amount approximately equal to the radius 1.22 (in units of λF) of the Airy disc.

The point with respect to which the variance of coma aberration is minimized is indicated by x_m (which from Section 8.3.3 is equal to $4FA_c/3$), and the irradiance at this point is given by I_m. We note that x_m and x_p are approximately equal to each other only for small values of $A_3(<0.7\lambda)$, showing that coma balanced with wavefront tilt to give minimum aberration variance across the pupil (i.e., Zernike coma) yields a maximum of irradiance only for small aberrations.

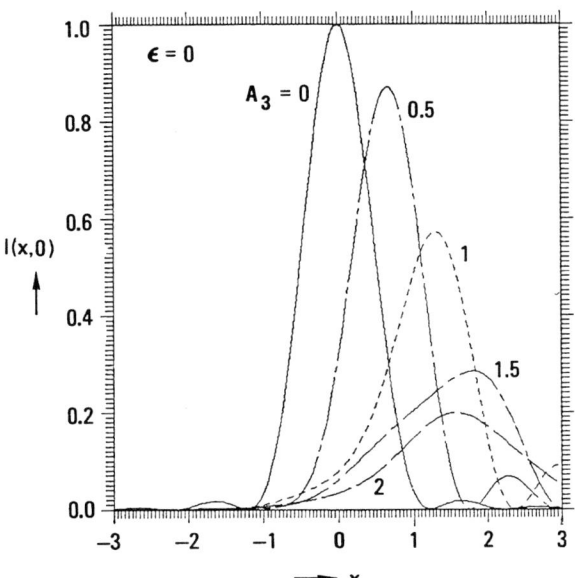

Figure 10–1. PSF profiles $I(x,0)$ normalized by the aberration-free central irradiance $I(0,0)$ for several values of coma aberration A_3 (from Reference 1).

Table 10–1. Typical values of x_m, x_p, and $<x>$ in units of λF and corresponding irradiances I_m, I_p, and I_c in units of the aberration-free central irradiance of PSFs for uniformly illuminated circular pupils aberrated by primary coma* (from Reference 1).

A_3	x_m	x_p	$<x>$	I_m	I_p	I_c	$I(0,0)$
0	0	0	0	1	1	1	1
0.5	0.67	0.66	1.00	0.8712	0.8712	0.6535	0.3175
1.0	1.33	1.30	2.00	0.5708	0.5717	0.1445	0.0791
1.5	2.00	1.80	3.00	0.2715	0.2844	0.0004	0.0618
2.0	2.67	1.57	4.00	0.0864	0.1978	0.0061	0.0341

*The units of A_3 are λ. The aberrated central irradiance $I(0,0)$ is also given here.

10.4 Comments

The results given here are applicable to both imaging systems, e.g., those used for optical surveillance, as well as to laser transmitters used for active illumination of a target. In both cases, the LOS of the optical system is extremely important. A LOS error of a surveillance system will produce an error in the location of the target. In the case of a laser transmitter, a large LOS error may cause the laser beam to miss the target altogether. Whereas for static aberrations we may be able to calibrate the LOS, for dynamic aberrations it is the analysis given here that will determine the tolerances of aberrations of the type $\rho^k\cos\theta$ and $\rho^k\sin\theta$.

Although we have defined the LOS of an optical system in terms of the centroid of its PSF, it could have been defined in terms of the peak of the PSF (assuming that the aberrations are small enough so that the PSF has a distinguishable peak). For an aberration–free PSF, its peak value and its centroid both lie at its origin, regardless of the amplitude variations across its pupil. The two are not coincident when $\cos\theta$ and/or $\sin\theta$ dependent aberrations are present. The precise definition of the LOS will perhaps depend on the nature of the application of the optical system. Moreover, in practice, only a finite central portion of the PSF will be sampled to measure its centroid, and the precision of this measurement will be limited by the noise characteristics of the photodetector array.

For simplicity, we have limited our discussion here to optical systems with uniform-circular pupils. However, the analysis can be extended to obtain the LOS errors of aberrated systems with *annular* and/or *Gaussian pupils*.[1] For example, for an annular pupil with a central obscuration ϵ, the right-hand side of Eq. (10-5a) is multiplied by $1 + \epsilon^2$ for $k = 3$. Compared to a uniform pupil, the value of $<x>$ for a Gaussian pupil is smaller; i.e., the centroid for a Gaussian pupil is closer to the true (aberration-free) LOS.

References

1.   V. N. Mahajan, "Line of sight of an aberrated optical system," J. Opt. Soc. Am. **A2**, 833-846 (1985).

CHAPTER 11

Random Aberrations

11.1 Introduction

So far we have considered *deterministic aberrations* such as those that are inherent in the design of an optical imaging system. These aberrations are deterministic in the sense that they are known or can be calculated, for example, by ray tracing the system. Now, we consider the effects of aberrations that are *random* in nature on the quality of images. The aberration is random in the sense that it varies randomly with time for a given system, or it varies randomly from one sample of a system to another. An example of the first kind is the aberration introduced by *atmospheric turbulence* when an optical wave propagates through it, as in ground-based astronomical observations. An example of the second kind is the aberration introduced due to *polishing errors* of the optical elements of the system. The polishing errors of an element fabricated similarly in large quantities vary randomly from one sample to another. In either case, we cannot obtain the exact image unless the instantaneous aberration or the exact polishing errors are known. However, based on the statistics of the aberrations, we can obtain the time- or ensemble-averaged image.

We discuss the effects of two types of random aberrations: *random wavefront tilt* causing *random image motion* and random aberrations introduced by atmospheric turbulence. The time-averaged Strehl ratio, point-spread function (PSF), optical transfer function (OTF), and encircled power are discussed for the two types of aberrations. Although much of our discussion is on systems with circular pupils, systems with annular pupils are also considered. A brief discussion on the aberrations resulting from *fabrication errors* is also given.

11.2 Theory

Consider an imaging system with a circular exit pupil of radius a imaging a point object radiating at a wavelength λ. Let $\Phi(\vec{r}_p)$ be the random phase aberration at a point $\vec{r}_p$ in the plane of the pupil. We assume that $\Phi(\vec{r}_p)$ is a stationary Gaussian random variable of zero mean, variance σ_Φ^2, and *structure function*

$$\mathfrak{D}_\Phi(|\vec{r}_p - \vec{r}_p'|) = <[\Phi(\vec{r}_p) - \Phi(\vec{r}_p')]^2>$$
$$= 2[\sigma_\Phi^2 - R_\Phi(|\vec{r}_p - \vec{r}_p'|)], \qquad (11\text{-}1)$$

where R_Φ is the *phase autocorrelation function* given by

$$R_\Phi(|\vec{r}_p - \vec{r}_p'|) = <\Phi(\vec{r}_p)\Phi(\vec{r}_p')> . \qquad (11\text{-}2)$$

Note that $R_\Phi(0) = \sigma_\Phi^2$ so that $\mathfrak{D}_\Phi(0) = 0$, as expected. The angular brackets in the above equations indicate a time or an ensemble average. For such an aberration, it can be shown that the *ensemble-* or *time-averaged* PSF, or the irradiance distribution of the image of a point object, normalized by its aberration-free central value, is given by

$$< I(r) > \; = \; 8 \int_0^1 <\tau(v)> J_0(2\pi r v)\,v\,dv \; , \tag{11-3}$$

where r is in units of λF (F being the focal ratio or the f-number of the image-forming light cone) and the *average* OTF corresponding to a spatial frequency v (in units of $1/\lambda F$) is related to the aberration-free OTF $\tau(v)$ according to

$$<\tau(v)> \; = \; \tau(v)\exp\left[-\frac{1}{2}\mathfrak{D}_\Phi(Dv)\right] . \tag{11-4}$$

Here, $D = 2a$ is the diameter of the exit pupil. The aberration-free OTF is given by (see Section 8.5)

$$\tau(v) = (2/\pi)[\cos^{-1} v - v(1-v^2)^{1/2}] \; , \; 0 \le v \le 1$$

$$= 0, \; \text{otherwise} . \tag{11-5}$$

The corresponding *average Strehl ratio* and *encircled power* are given by

$$<S> \; = \; <I(0)>$$

$$= 8 \int_0^1 <\tau(v)>\,v\,dv \tag{11-6}$$

and

$$<P(r_c)> \; = \; 2\pi r_c \int_0^1 <\tau(v)> J_1(2\pi r_c v)\,dv \; , \tag{11-7}$$

respectively. Now we apply these results to determine the effects of random image motion and random aberrations introduced by atmospheric turbulence on the OTF, PSF, Strehl ratio, and encircled power of a system.

11.3 Random Image Motion

In many optical imaging systems, especially those used in space, there is always some *image motion* during an exposure interval. The source of image motion may, for example, be vibration of optical elements and servo dither in the pointing system. In the case of beam-transmitting systems, the beam itself may have some motion associated with it.

It should be evident that an image motion is equivalent to a time-dependent *wavefront tilt*. Thus, for example, a wavefront tilt of a small angle β about the y axis is equivalent to a linear motion of $R\beta$ along the x axis in the image plane, where R is the distance between the pupil plane and the observation plane [see Eq. (1-4) and Figure 1-1]. The corresponding wave aberration at a point $(\vec{r}_p, \theta)$ in the plane of the exit pupil is $\beta r_p \cos\theta$ or βx_p. For a system introducing a two-dimensional image motion, the corresponding wave aberration may be written

$$W(\vec{r}_p; \beta) = \vec{\beta} \cdot \vec{r}_p \ , \tag{11-8}$$

where $\vec{\beta}$ represents the two-dimensional wavefront tilt or angular motion of the image. Following Eq. (11-1), the resultant *phase structure function* is given by

$$\mathcal{D}_\Phi(|\vec{r}_p - \vec{r}_p'|) = (2\pi/\lambda)^2 < [\vec{\beta} \cdot (\vec{r}_p - \vec{r}_p')]^2 > \tag{11-9}$$
$$= 4\pi^2 < (R\vec{\beta} \cdot \vec{v}_i)^2 > \ ,$$

where $\vec{v}_i = (\vec{r}_p - \vec{r}_p')/\lambda R$ is a spatial frequency. If the image motion is described by statistically independent Gaussian random processes of zero mean and equal standard deviation σ, in units of λF, along the two axes of the image, then Eq. (11-9) reduces to

$$\mathcal{D}_\Phi(|\vec{r}_p - \vec{r}_p'|) = 4\pi^2 \sigma^2 v^2 \ , \tag{11-10}$$

where $v = \lambda F |\vec{v}_i|$ is a spatial frequency in units of $1/\lambda F$. Substituting Eq. (11-10) into Eq. (11-4), we obtain the time-averaged OTF of the system:

$$< \tau(v) > = \tau(v) \exp(-2\pi^2 \sigma^2 v^2) \ . \tag{11-11}$$

Since the PSF and the OTF of a system form a Fourier transform pair, Eq. (11-11), when Fourier transformed to give the time-averaged PSF, represents a convolution of the motion-free PSF and the Gaussian probability density function describing its motion.[1] Substituting Eq. (11-11) into Eqs. (11-3), (11-6), and (11-7), we can calculate the time-averaged PSF, Strehl ratio, and encircled power, respectively. Equation (11-6) for the Strehl ratio yields

$$< S(\sigma) > = (2/\pi^2\sigma^2)\{1 - \exp(-\pi^2\sigma^2)[I_0(\pi^2\sigma^2) + I_1(\pi^2\sigma^2)]\} \ , \tag{11-12}$$

where $I_0(\bullet)$ and $I_1(\bullet)$ are the hyperbolic or modified Bessel functions of zero and first order, respectively.

The *Gaussian approximation* of the motion-free PSF, having the same central value and total power as the actual PSF, i.e., the *Airy pattern*, is given by

$$I_g(r) = \exp[-(\pi r/2)^2] \ . \tag{11-13}$$

The corresponding approximation of the OTF is given by

$$\tau_g(r) = \exp(-4\,v^2)\ . \tag{11-14}$$

For a *Gaussian random motion* as considered above, the time-averaged PSF is also a Gaussian with a variance equal to the sum of the variances of the motion-free PSF and the image motion, i.e.,

$$<I_g\,(r;\sigma)> \ = \ [(\pi^2/2)(\sigma^2 + 2/\pi^2)]^{-1}\exp[-r^2/2(\sigma^2 + 2/\pi^2)]\ . \tag{11-15}$$

The corresponding time-averaged OTF is given by

$$<\tau_g\,(v\,;\sigma)> \ = \ \exp[-2\pi^2\,(\sigma^2 + 2/\pi^2)v^2]\ . \tag{11-16}$$

The time-averaged Strehl ratio and encircled power are given by

$$<S_g\,(\sigma)> \ = \ (1 + \pi^2\sigma^2/2)^{-1} \tag{11-17}$$

and

$$<P_g\,(r_c;\sigma)> \ = \ 1 - \exp[-r_c^2/2(\sigma^2 + 2/\pi^2)]\ , \tag{11-18}$$

respectively. For large values of σ, Eqs. (11-12) and (11-17) may be written

$$<S_g\,(\sigma)> \ \rightarrow \ 2/\pi^2\sigma^2\ . \tag{11-19}$$

Figure 11-1 shows how the time-averaged Strehl ratio varies with σ according to Eqs. (11-12) and (11-17). The exact results are shown by the solid curves and the corresponding approximate results are shown by the dashed curves. We note that the Gaussian approximation given by Eq. (11-17) overestimates the Strehl ratio. However, the maximum difference $(S - S_g)/S$ is less than 12% and occurs at $\sigma = 1.1$. The time-averaged PSFs obtained according to Eqs. (11-3) and (11-15) for typical values of σ are shown in Figure 11-2. The corresponding encircled powers obtained by use of Eqs. (11-7) and (11-18) are shown in Figure 11-3. Once again, we note that the Gaussian approximation overestimates the irradiance (at least *within* the Airy disc) and the encircled power. However, the differences between the exact and approximate results are not large. In the case of encircled power, the differences increase monotonically as the radius of the circle increases.

11.4 Atmospheric Turbulence

The phase structure function for an optical wave propagating through *Kolmogorov atmospheric turbulence* may be written[2]

$$\mathcal{D}_\Phi(\vec{r}_p - \vec{r}_p') = 6.88\left(\frac{|\vec{r}_p - \vec{r}_p'|}{r_0}\right)^{5/3}\ , \tag{11-20}$$

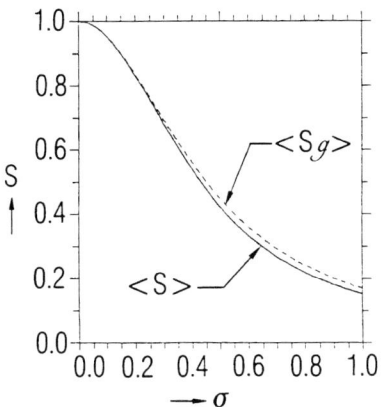
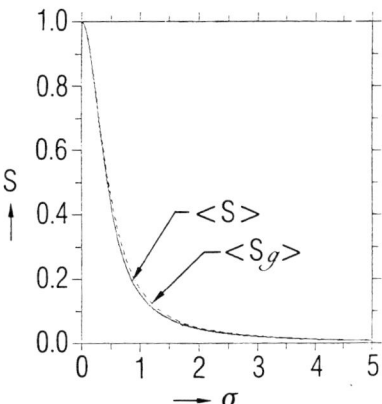

Figure 11–1. Average Strehl ratio for Gaussian image motion characterized by its standard deviation σ in units of λF along each of the two axes of an image.

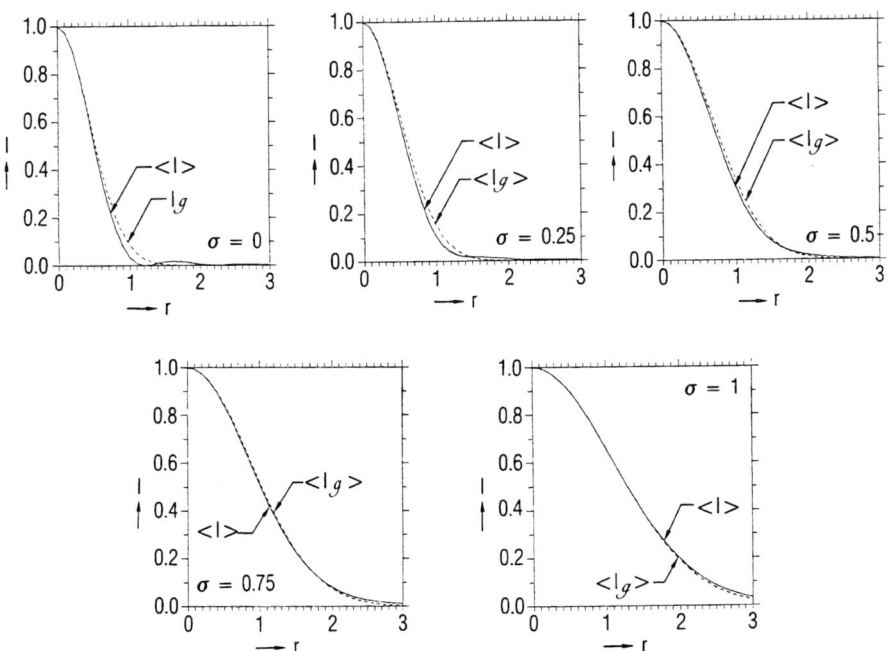

Figure 11–2. Average irradiance distribution for Gaussian image motion. r is in units of λF.

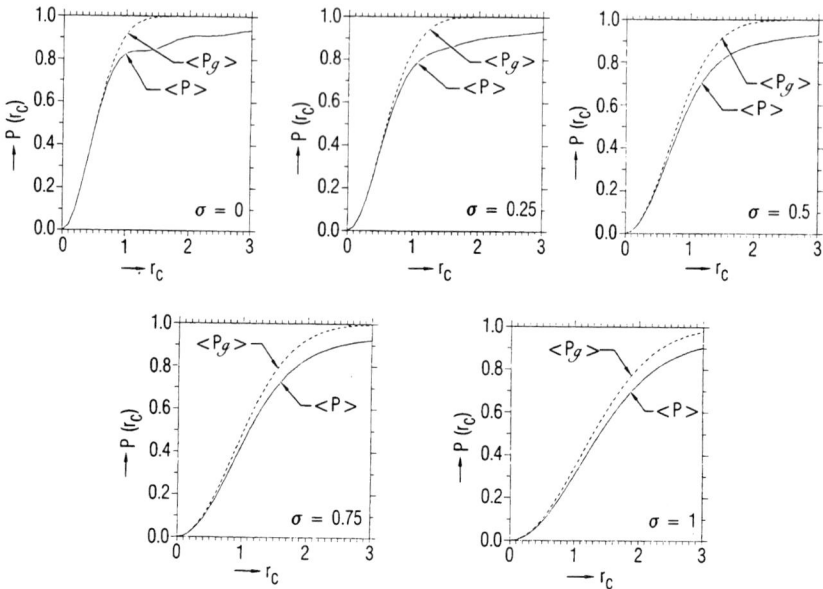

Figure 11–3. Average encircled power for Gaussian image motion. r_c is in units of λF.

where r_0 is a characteristic length of turbulence called its *coherence length* or *diameter*. Substituting Eq. (11-20) into Eq. (11-4) we may write

$$< \tau (v) > \ = \ \tau(v)\tau_a(v) \tag{11-21}$$

where

$$\tau_a(v) \ = \ \exp[-3.44(D \, v/r_0)^{5/3}] \tag{11-22}$$

is called the atmospheric OTF. We note that, since $\exp(-3.44) \approx 0.03$, atmospheric turbulence reduces the system OTF corresponding to a spatial frequency $v = r_0/D$ by a factor of 0.03. Similarly, the *degree of coherence* of complex amplitudes at two points on a wave propagating through turbulence that are separated by a distance r_0 is 0.03, or that the *visibility of the fringes* formed by an interference of the secondary waves from these points is 0.03.

Substituting Eq. (11-21) into Eqs. (11-3), (11-6), and (11-7), we can evaluate the average irradiance distribution, Strehl ratio, and encircled power, respectively. Figure 11-4 shows how $(D/r_0)^2 < S >$ which is proportional to the aberrated central irradiance, varies with D/r_0. For small values of D, the central irradiance increases as D^2 as for an aberration-free system. As D increases, the central irradiance increases much more slowly than D^2, and for $D/r_0 > 5$, the increase is very small. As $D/r_0 \to \infty$, the central irradiance approaches a value of PS_a/λ^2R^2, where $S_a = (\pi/4)r_0^2$ is the coherence area

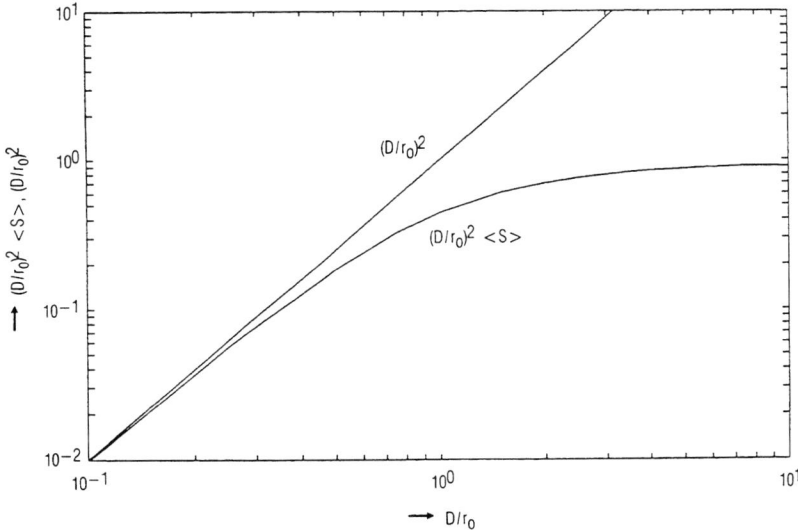

Figure 11–4. Variation of $(D/r_0)^2 <S>$ with D/r_0.

of the atmosphere. This corresponds to the aberration-free value for a system with an exit pupil of diameter r_0.

In astronomical observations, the total power at the exit pupil $P = (\pi/4) D^2 I_0$, where I_0 is the average irradiance across the exit pupil, increases as D increases. However, if the observation is made against a *uniform background*, then the background irradiance in the image also increases as D^2. Hence, the detectability of a point object is limited by turbulence to a value corresponding to an exit pupil of diameter r_0, no matter how large the diameter D of the exit pupil is. In the case of a laser transmitter with a fixed value of laser power, the central irradiance on a target will again be limited to its aberration-free value for an exit pupil of diameter r_0, no matter how large the transmitter diameter is.

Figure 11-5 shows how $<S>$ varies with D/r_0. It decreases to zero monotonically as D/r_0 increases. Thus, for example, for a given value of D, the Strehl ratio decreases rapidly as r_0 decreases. Even when r_0 is as large as D, the Strehl ratio is only 0.445. Its value varies with the wavelength of object radiation as $\lambda^{1.2}$. At a visible wavelength, its approximate value is only 10 cm. Accordingly, the performance of a ground-based telescope making astronomical observations in the visible region is limited primarily by atmospheric turbulence. Figure 11-6 shows the average irradiance distribution for several values of D/r_0. The corresponding encircled-power distribution is also shown in this figure. As D/r_0 increases, a given fraction of the total power is contained in a larger and larger circle. As an example, whereas 84 percent of the total power is

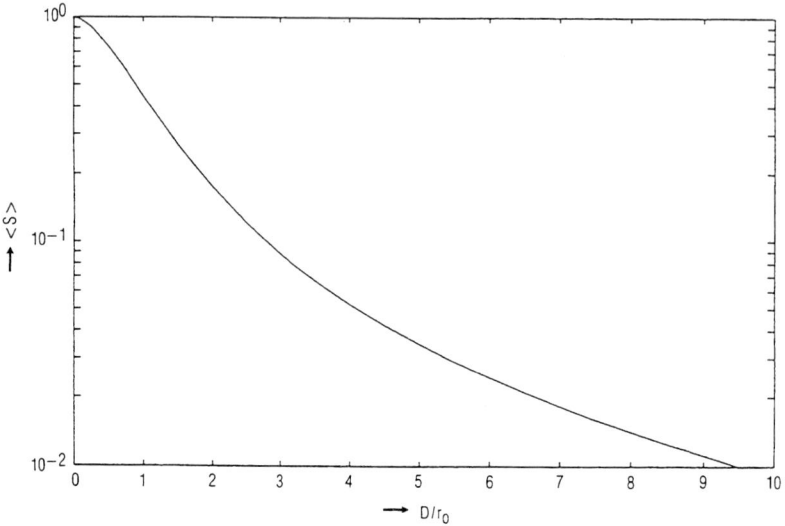

Figure 11–5. Variation of average Strehl ratio $<S>$ with D/r_0.

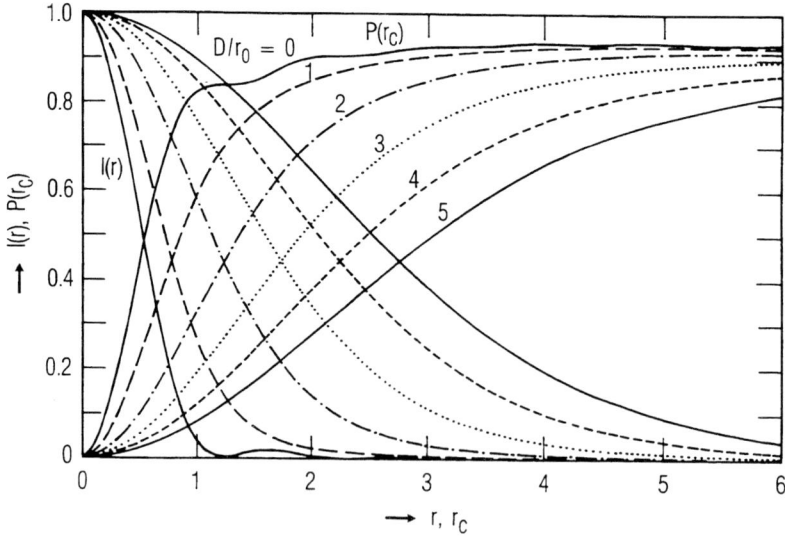

Figure 11–6. Average irradiance and encircled-power distributions for different values of D/r_0. r and r_c are in units of λF.

contained in a circle of radius r_c = 1.22 when there is no turbulence, it is contained in a circle of radius 1.9 when D/r_0 = 1.

The time-averaged variance of the phase aberration introduced by Kolmogorov turbulence is given by[3]

$$\sigma_\Phi^2 = 1.03(D/r_0)^{5/3} \ .$$ (11-23)

The aberration may be decomposed into various aberration types in terms of Zernike polynomials. It is found, for example, that much of the aberration consists of wavefront tilt, i.e., random image motion. The aberration variance of a short-exposure image (which is not degraded by its motion) is only 0.134 $(D/r_0)^{5/3}$.

Figure 11-7 shows short-exposure PSFs corresponding to different values of D/r_0. In Part (a), D is kept fixed while r_0 decreases from a value equal to D to $D/3$ and to $D/10$, e.g., D = 1m, r_0 = 1m, 33.3 cm, and 10 cm, respectively. We note that each image is broken up into small spots called *speckles*, which is a characteristic of random aberrations. The size of a speckle is determined by D, its angular radius being approximately equal to λ/D. The size of the total image is determined by r_0, its angular radius being approximately equal to λ/r_0. The image becomes progressively worse as r_0 decreases showing the effects of what astronomers call *seeing*. In Part (b), the value of r_0 is kept fixed while the value of D increases from a value equal to r_0 to $3r_0$ and to $10r_0$, e.g., r_0 = 10 cm, D = 10 cm, 30 cm, and 1 m, respectively. Now the size of a speckle decreases as D increases, but the image size is approximately constant. Thus, an increase in D does not significantly improve the resolution of the system (as determined by the total size of the image). For convenience, the pictures in Part (b) are shown reduced by a factor of 2.5 compared to those in Part (a). Thus, for example, the pictures corresponding to D/r_0 = 10 in these two parts are otherwise similar. (The aberration function used for this case and the corresponding interferogram are shown in Figure 12-4.)

The approximate expressions of Eqs. (8-13)–(8-15) are not suitable for calculating the average Strehl ratios for random aberrations. For example, even for D/r_0 = 1, Eq. (8-15) gives a Strehl ratio of 0.357, compared to a true value of 0.445. For larger values of D/r_0, Eq. (8-15) underestimates the average Strehl ratio by larger factors.

11.5 Annular Pupils

For optical systems with annular pupils, Eqs. (11-3), (11-6), and (11-7) for the average PSF, Strehl ratio, and encircled power, respectively, can be used provided $\tau(v)$ is replaced by the aberration-free OTF $\tau(v;\epsilon)$ for an annular pupil, where ϵ is the obscuration ratio of the pupil. The average Strehl ratio thus obtained in the case of a Gaussian random motion of the image is given in Table 11-1. Its variation with image motion is also shown in Figure 11-8 for a few values of ϵ. We note that as the obscuration of the pupil increases, the drop in Strehl ratio due to image motion for a given value of σ also increases.

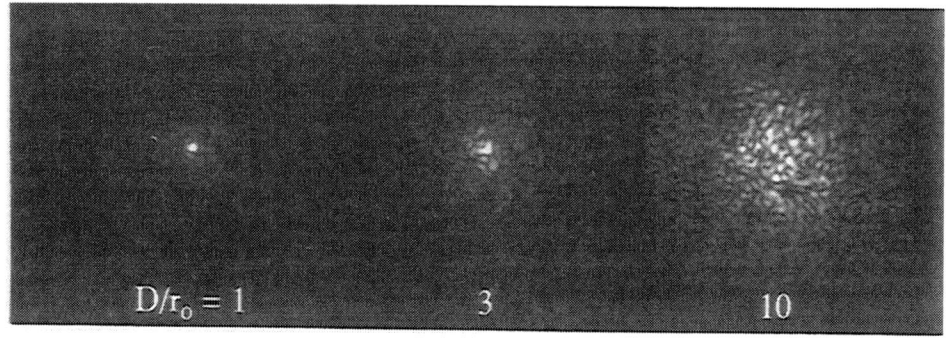

(a)

(b)

Figure 11–7. Short-exposure PSFs aberrated by atmospheric turbulence. In (a), D is kept fixed and r_0 is varied. In (b), r_0 is kept fixed and D is varied. The value of D determines the size of a speckle while r_0 determines the image size. For convenience, pictures in (b) are shown reduced by a factor of 2.5 compared to those in (a).

Table 11-1. Time-averaged Strehl ratio for Gaussian image motion characterized by σ in units of λF (from Reference 1).

σ\ ϵ	0.00	0.05	0.10	0.15	0.20	0.25	0.30	0.35	0.40	0.45	0.50	0.55	0.60	0.65	0.70	0.75	0.80	0.85	0.90	0.95
0	1	1	1	1	1	1	1	1	1	1	1	1	1	1	1	1	1	1	1	1
0.05	0.988	0.988	0.988	0.987	0.987	0.987	0.987	0.986	0.986	0.985	0.985	0.984	0.983	0.983	0.982	0.981	0.980	0.979	0.979	0.981
0.10	0.953	0.952	0.952	0.952	0.951	0.950	0.948	0.947	0.945	0.943	0.941	0.939	0.936	0.933	0.930	0.927	0.924	0.920	0.917	0.916
0.15	0.898	0.898	0.898	0.896	0.895	0.892	0.890	0.887	0.883	0.879	0.875	0.870	0.865	0.859	0.853	0.847	0.841	0.834	0.827	0.823
0.20	0.831	0.831	0.830	0.828	0.825	0.822	0.817	0.813	0.807	0.801	0.794	0.787	0.779	0.771	0.762	0.752	0.743	0.733	0.723	0.715
0.25	0.756	0.756	0.754	0.752	0.748	0.744	0.738	0.732	0.724	0.716	0.708	0.698	0.688	0.677	0.666	0.655	0.643	0.630	0.618	0.608
0.30	0.680	0.679	0.678	0.674	0.670	0.665	0.658	0.650	0.642	0.632	0.622	0.611	0.599	0.587	0.575	0.562	0.549	0.536	0.524	0.513
0.35	0.606	0.606	0.603	0.600	0.595	0.589	0.582	0.573	0.564	0.554	0.543	0.531	0.519	0.507	0.494	0.481	0.468	0.456	0.443	0.433
0.40	0.538	0.537	0.535	0.531	0.526	0.520	0.512	0.503	0.494	0.483	0.472	0.461	0.449	0.437	0.425	0.413	0.402	0.390	0.379	0.370
0.45	0.476	0.476	0.473	0.469	0.464	0.458	0.450	0.442	0.433	0.423	0.412	0.402	0.391	0.380	0.369	0.359	0.348	0.338	0.328	0.321
0.50	0.422	0.421	0.419	0.415	0.410	0.404	0.397	0.389	0.380	0.371	0.362	0.352	0.343	0.333	0.324	0.315	0.306	0.297	0.289	0.283
0.55	0.375	0.374	0.372	0.368	0.364	0.358	0.351	0.344	0.336	0.328	0.320	0.312	0.304	0.296	0.288	0.280	0.273	0.265	0.258	0.253
0.60	0.334	0.333	0.331	0.328	0.323	0.318	0.312	0.306	0.299	0.293	0.286	0.279	0.272	0.265	0.259	0.252	0.246	0.240	0.234	0.229
0.65	0.298	0.298	0.296	0.293	0.289	0.285	0.279	0.274	0.268	0.263	0.257	0.251	0.246	0.240	0.235	0.229	0.224	0.218	0.212	0.203
0.70	0.268	0.267	0.265	0.263	0.260	0.256	0.251	0.247	0.242	0.237	0.233	0.228	0.224	0.219	0.215	0.210	0.206	0.201	0.196	0.187
0.75	0.241	0.241	0.239	0.237	0.234	0.231	0.227	0.223	0.220	0.216	0.212	0.209	0.205	0.202	0.198	0.194	0.190	0.186	0.181	0.174
0.80	0.218	0.218	0.216	0.215	0.212	0.209	0.206	0.203	0.201	0.198	0.195	0.192	0.189	0.187	0.184	0.181	0.177	0.174	0.169	0.163
0.85	0.198	0.198	0.197	0.195	0.193	0.191	0.188	0.186	0.184	0.182	0.180	0.178	0.176	0.173	0.171	0.169	0.166	0.163	0.159	0.153
0.90	0.181	0.180	0.179	0.178	0.176	0.175	0.173	0.171	0.169	0.168	0.166	0.165	0.164	0.162	0.160	0.158	0.156	0.153	0.150	0.144
0.95	0.165	0.165	0.164	0.163	0.162	0.160	0.159	0.158	0.157	0.156	0.155	0.154	0.153	0.152	0.150	0.149	0.147	0.144	0.141	0.136
1.00	0.152	0.152	0.151	0.150	0.149	0.148	0.147	0.146	0.145	0.145	0.144	0.144	0.143	0.143	0.142	0.140	0.139	0.137	0.134	0.131

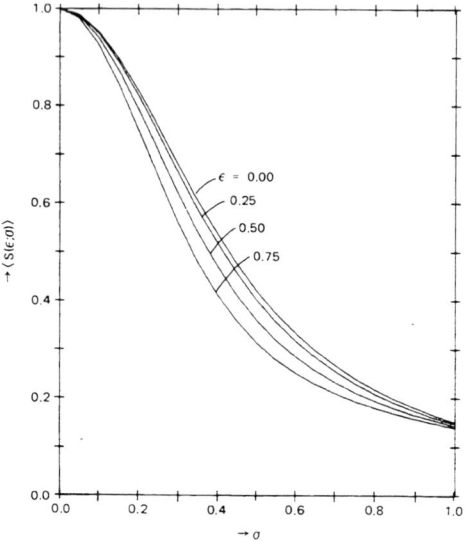

Figure 11–8. Average Strehl ratio as a function of σ for several values of ϵ (from Reference 1).

Table 11-2 gives the average Strehl ratio in the case of imaging through Kolmogorov atmospheric turbulence for several values of ϵ and D/r_0. Figure 11-9 shows how a quantity

$$\eta(D/r_0; \epsilon) = (1 - \epsilon^2)(D/r_0)^2 < S\,(r_0; \epsilon)> \quad , \tag{11-24}$$

which is proportional to the aberrated central irradiance, varies with D/r_0 for a given value of ϵ and a fixed total power P. For small values of D, the central irradiance increases as D^2 as for an aberration-free system. However, because of turbulence, it does not increase significantly with D beyond a certain value of D/r_0. The saturation effects of atmospheric turbulence occur at larger and larger values of D/r_0 as ϵ increases. We note that irrespective of the value of ϵ

$$\eta(D/r_0; \epsilon) \to 1 \ as \ D \to \infty \quad . \tag{11-25}$$

Table 11–2. Average Strehl ratio for various values of ϵ and D/r_0 (from Reference 4).

$\epsilon \backslash D/r_0$	1	2	3	4	5
0	0.445	0.175	0.089	0.053	0.035
0.25	0.430	0.169	0.088	0.054	0.036
0.50	0.391	0.160	0.090	0.058	0.040
0.75	0.344	0.152	0.095	0.067	0.050

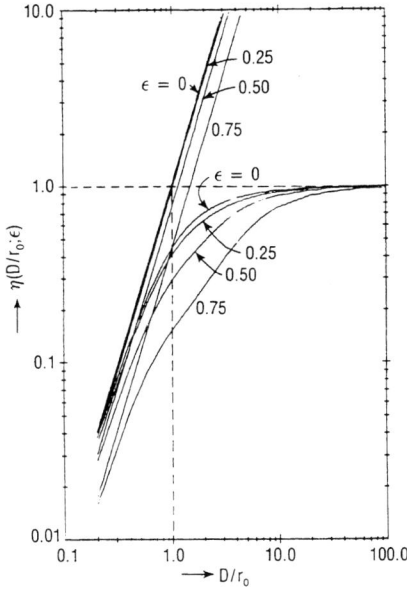

Figure 11-9. Variation of $\eta(D/r_0;\epsilon) = (1 - \epsilon^2)(D/r_0)^2 < S(r_0;\epsilon) >$ with D/r_0 (from Reference 4).

Moreover,

$$\eta(D/r_0;\epsilon) \rightarrow (1 - \epsilon^2)(D/r_0)^2 \quad as \quad D \rightarrow 0 . \tag{11-26}$$

The two asymptotes of $\eta(D/r_0;\epsilon)$ intersect at $D/r_0 = (1 - \epsilon^2)^{1/2}$. The central irradiance is given by $(PS_p/\lambda^2R^2) < S(r_0;\epsilon >$ or $(PS_a/\lambda^2R^2)\eta(D/r_0;\epsilon)$, where $S_p = \pi(1 - \epsilon^2) a^2$ is the area of the annular pupil and $S_a = \pi r_0^2/4$ is the coherence area of the atmosphere. Hence, regardless of how large D is, the central irradiance is less than or equal to the aberration-free central irradiance for a system with an exit pupil of diameter r_0, equality approaching as $D \rightarrow \infty$. The limiting value of the central irradiance is independent of the value of ϵ.

Figure 11-10 shows how the encircled power in the Airy disc, i.e., for $r_c = 1.22$, varies with D/r_0. It is evident from this figure that for a given value of D/r_0, the relative loss of power from the Airy disc decreases as ϵ increases.

11.6 Fabrication Errors

In Chapters 1-6, we have shown how to calculate the aberrations of various optical imaging systems. Although it was not pointed out explicitly, it was understood that the elements of a system had their *prescribed shapes*, i.e., the elements did not have any fabrication errors. The aberrations of a system thus calculated are referred to as its

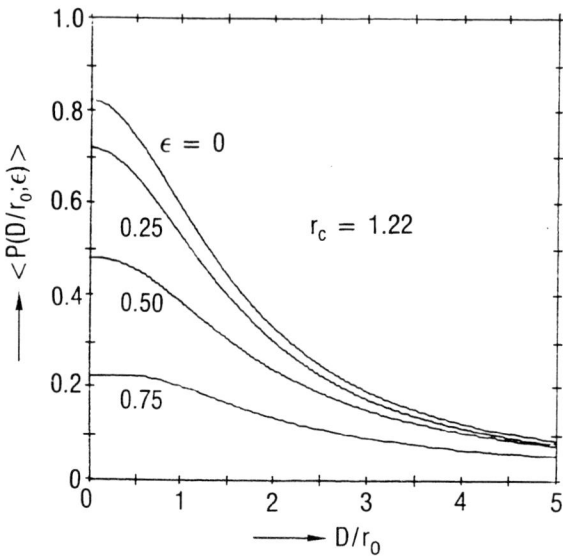

Figure 11–10. Encircled power in Airy disc as a function of D/r_0 for several values of ϵ (from Reference 4).

design aberrations. In practice, when the elements of a system are fabricated, their exact shapes will deviate however slightly from their prescribed shapes. These *fabrication* or *manufacturing errors* are generally referred to as their *surface* or *figure errors*. They are typically random in nature in that if an element is fabricated in large quantities, its errors will vary randomly from one sample to another. However, these errors have certain statistical properties which depend on the fabrication process. For example, the width (*correlation length*) of the polishing irregularities of an element depends on the size of the tool used to polish it.

The figure errors of an element of a system contribute to its aberrations. For example, if θ and θ' are the angles of incidence and refraction of a ray incident on a refracting surface separating media of refractive indices n and n', and if δF is the deviation of the surface at the point of incidence of the ray along the surface normal at that point from the prescribed shape, the change in its optical path length is given by

$$\delta W = (n\cos\theta - n'\cos\theta')\delta F . \tag{11-27}$$

Thus, under normal incidence, a plane–parallel plate of refractive index n introduces wavefront errors that are $(n-1)$ times its corresponding figure errors. The figure errors of its two surfaces are added in a *root sum square sense* as mentioned below. In the case of a reflecting surface in air, Eq. (11-27) reduces to

$$\delta W = 2\cos\theta\delta F \ . \tag{11-28}$$

Thus, a conservative estimate of the wavefront errors in this case is equal to twice the figure errors. The wavefront errors arising from *thermal distortions* of the elements and their *misalignments* and *spacing errors* may also be calculated by using Eqs. (11-27) and (11-28).

If a system is tested and its aberrations are measured, its imaging properties can be calculated by the techniques outlined in Section 8.2. However, if the phase aberrations of a system based on its few samples obey Gaussian statistics as assumed in Section 11.2, the expected value, for example, of the PSF of another sample can be obtained by the use of Eq. (11-3).

Because of the random nature of the figure errors, the total expected wavefront error of the system can be obtained by a square root of the sum of the variances of the wavefront errors contributed by its elements. Indeed this is how optical tolerances on the figure errors of the elements of a system are allocated. For example, if we are interested in a system Strehl ratio of 0.8 so that the total budget for the *standard deviation* of the wavefront errors is $\lambda/14$, the figure errors of the elements can be allocated equally or preferentially among them such that the root sum square of the standard deviations of their wavefront errors is $\lambda/14$.

References

1.   V. N. Mahajan, "Degradation of an image due to Gaussian image motion," Appl. Opt. **17**, 3329–3334 (1978).

2.   D. Fried, "Optical heterodyne detection of an atmospherically distorted signal wave front," Proc. IEEE **55**, 57–67 (1967).

3.   R. J. Noll, "Zernike polynomials and atmospheric turbulence," J. Opt. Soc. Am. **66**, 207–211 (1976).

4.   V. N. Mahajan and B. K. C. Lum, "Imaging through atmospheric turbulence with annular pupils," Appl. Opt. **20**, 3233–3237 (1981).

CHAPTER 12

Observation of Aberrations

12.1 Introduction

In this chapter, we describe briefly how the primary aberrations of an optical system can be observed. The emphasis of our discussion is on how to *recognize a primary aberration and not on how to measure it precisely.* Since the optical frequencies are very high ($10^{14} - 10^{15}$ Hz), optical wavefronts, aberrated or not, cannot be observed directly; optical detectors simply do not respond at these frequencies. We have seen in Chapter 8 that the image of a monochromatic point object formed by an aberrated system is characteristically different for a different aberration. Another and more powerful way to recognize an aberration is to form an *interferogram* by combining two parts of a light beam, one of which has been transmitted through the system. An aberration in the system yields an *interference pattern* that is characteristically different for a different aberration. Here, we briefly discuss the interference patterns for primary aberrations.

12.2 Primary Aberrations

Considering an optical system with a circular exit pupil of radius a and letting (r,θ) be the polar coordinates of a point in the plane of its exit pupil, the functional form of the *primary phase aberrations* may be written

$$\Phi(\rho,\theta) = \begin{cases} A_s\rho^4 + A_d\rho^2 & \text{Spherical combined with defocus} & (12\text{-}1) \\[2mm] A_c\rho^3\cos\theta + A_t\rho\cos\theta & \text{Coma combined with tilt} & (12\text{-}2) \\[2mm] A_a\rho^2\cos^2\theta + A_d\rho^2 & \text{Astigmatism combined with defocus} & (12\text{-}3) \\[2mm] A_d\rho^2 & \text{Defocus or field curvature} & (12\text{-}4) \\[2mm] A_t\rho\cos\theta & \text{Tilt or distortion,} & (12\text{-}5) \end{cases}$$

where, as in Section 1.6, A_i is a peak aberration coefficient representing the maximum value of the corresponding aberration across the pupil and $\rho = r/a$ is a normalized radial variable. When $\Phi(\rho,\theta) = 0$, the wavefront passing through the center of the exit pupil, for a point object, is spherical centered at the Gaussian image point. Let its radius of curvature be R. For an aberrated system, $\Phi(\rho,\theta)$ represents the optical deviation of the wavefront from being spherical at a point (ρ,θ).

In Eq. (12-1), when $A_d = 0$, the aberration is *spherical*. Nonzero A_d implies that the aberration is combined with defocus; i.e., the aberration is not with respect to a reference sphere centered at the Gaussian image point but with respect to another sphere centered at a distance z from the plane of the exit pupil according to Eq. (8-6). As discussed in Chapter 7, the reference sphere is centered at the marginal image point, center of the circle of least confusion, and the point midway between the marginal and Gaussian image points when A_d/A_s = -2, - 1.5, and -1, respectively. The midway point corresponds to minimum variance of the aberration and, therefore, to maximum Strehl ratio (for small aberrations).

In Eq. (12-2), when A_t = 0, the aberration is *coma*. Nonzero A_t implies that the aberration is combined with tilt, or that it is with respect to a reference sphere centered at a point $(2FA_t,0)$ in the image plane, where F is the focal ratio or the f–number of the image–forming light cone. The variance of the aberration is minimum when A_t/A_c = -2/3.

In Eq. (12-3), when A_d = 0, the aberration is *astigmatism*. Nonzero A_d implies that it is combined with defocus. The variance of the aberration is minimum when A_d/A_a = -1/2. When A_d/A_a = 0 or -1, we obtain the so-called tangential and sagittal images of a point object. Equations (12-4) and (12-5) represent defocus or field curvature and tilt or distortion aberrations, respectively. Figure 12-1 shows a 3-D plot of the various aberrations.

12.3 Interferograms

There are a variety of interferometers that are used for detecting and measuring aberrations of optical systems.[1] Figure 12-2 illustrates schematically a *Twyman-Green interferometer* in which a collimated laser beam is divided into two parts by a beam splitter *BS*. One part, called the *test beam*, is incident on the system under test, indicated by the lens *L*, and the other, called the *reference beam*, is incident on a plane mirror M_1. The focus *F* of the lens system lies at the center of curvature *C* of a spherical mirror M_2. As the angle of the incident light is changed to study the off-axis aberrations of the system, the mirror is tilted so that its center of curvature lies at the current focus of the beam. In this arrangement the mirror does not introduce any aberration since it is forming the image of an object lying at its center of curvature (see Section 4.2). The two reflected beams interfere in the region of their overlap. Lens *L'* is used to observe the interference pattern on a screen *S* placed in a plane containing the image of *L* formed by *L'*. A record of the interference pattern is called an *interferogram*. Note that since the test beam goes through the lens system *L* twice, its aberration is twice that of the system.

If the reference beam has a uniform phase and the test beam has a phase distribution $\Phi(x,y)$, and if their amplitudes are equal to each other, the irradiance distribution of their interference pattern is given by

$$I(x,y) = I_0|1 + \exp[i\Phi(x,y)]|^2$$

$$= 2I_0\{1 + \cos[\Phi(x,y)]\} \ , \tag{12-6}$$

where I_0 is the irradiance when only one beam is present. The irradiance has a maximum value equal to $4I_0$ at those points for which

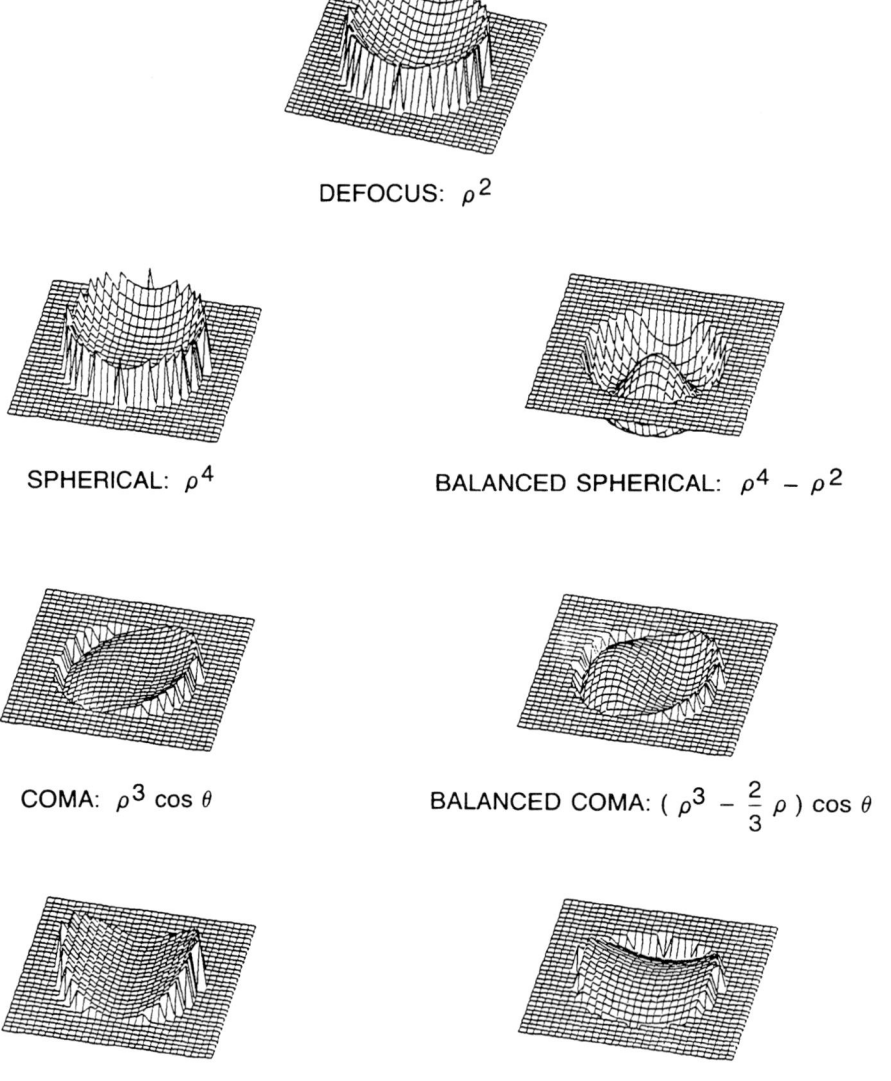

DEFOCUS: ρ^2

SPHERICAL: ρ^4

BALANCED SPHERICAL: $\rho^4 - \rho^2$

COMA: $\rho^3 \cos\theta$

BALANCED COMA: $(\rho^3 - \dfrac{2}{3}\rho)\cos\theta$

ASTIGMATISM: $\rho^2 \cos^2\theta$

BALANCED ASTIGMATISM: $\rho^2 \cos^2\theta - \dfrac{1}{2}\rho^2$

Figure 12–1. Shape of primary aberrations representing the difference between an ideal wavefront (typically, spherical) and an actual wavefront.

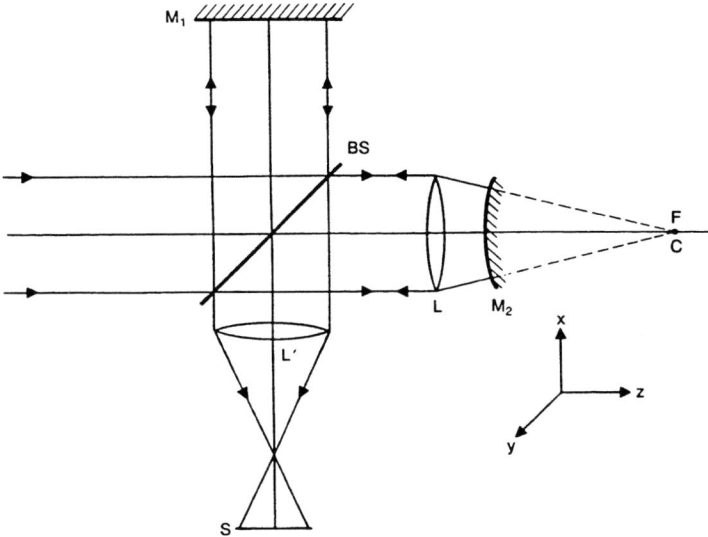

Figure 12-2. Twyman-Green interferometer for testing a lens system L.

$$\Phi(x,y) = 2\pi n \qquad\qquad (12\text{-}7a)$$

and a minimum value equal to zero wherever

$$\Phi(x,y) = 2\pi(n + 1/2) \; , \qquad\qquad (12\text{-}7b)$$

where n is a positive or a negative integer, including zero. Each fringe in the interference pattern represents a certain value of n, which in turn corresponds to the locus of (x,y) points with phase aberration given by Eq. (12–7a) for a bright fringe and Eq. (12–7b) for a dark fringe. If the test beam is aberration free [$\Phi(x,y) = 0$], then the interference pattern has a uniform irradiance of $2I_0$.

Figure 12-3 shows the interferograms when the lens system L under test suffers from 3λ of a primary aberration, corresponding to 6λ of an aberration of the interfering test beam. In our discussion, we give the value of an aberration coefficient in wavelength units, rather than in radians, as is customary in optics. For defocus and spherical aberration, the interference pattern consists of concentric circular interference fringes. The fringe spacing depends on the type of the aberration. Figure 12-3a shows the interferogram obtained when the system is aberration free but it is misfocused, i.e., when its focus F lies to the left or the right of the center of curvature C of the spherical mirror M_2 by an amount corresponding to 3λ of defocus aberration. [See Eqs. (1-3) for a relationship between the longitudinal defocus, i.e., the axial spacing between F and C, and the peak defocus aberration A_d, which is 3λ in our example.] Figure 12-3b shows the interferograms obtained when the system has 3λ of spherical aberration (i.e., $A_s = 3\lambda$) and a certain amount of defocus. The case $A_d = 0$ (i.e., F and C coincident) represents such a system with an image of a certain object being observed in its Gaussian or paraxial image plane. Similarly, the interferogram obtained for $A_d/A_s = -2$

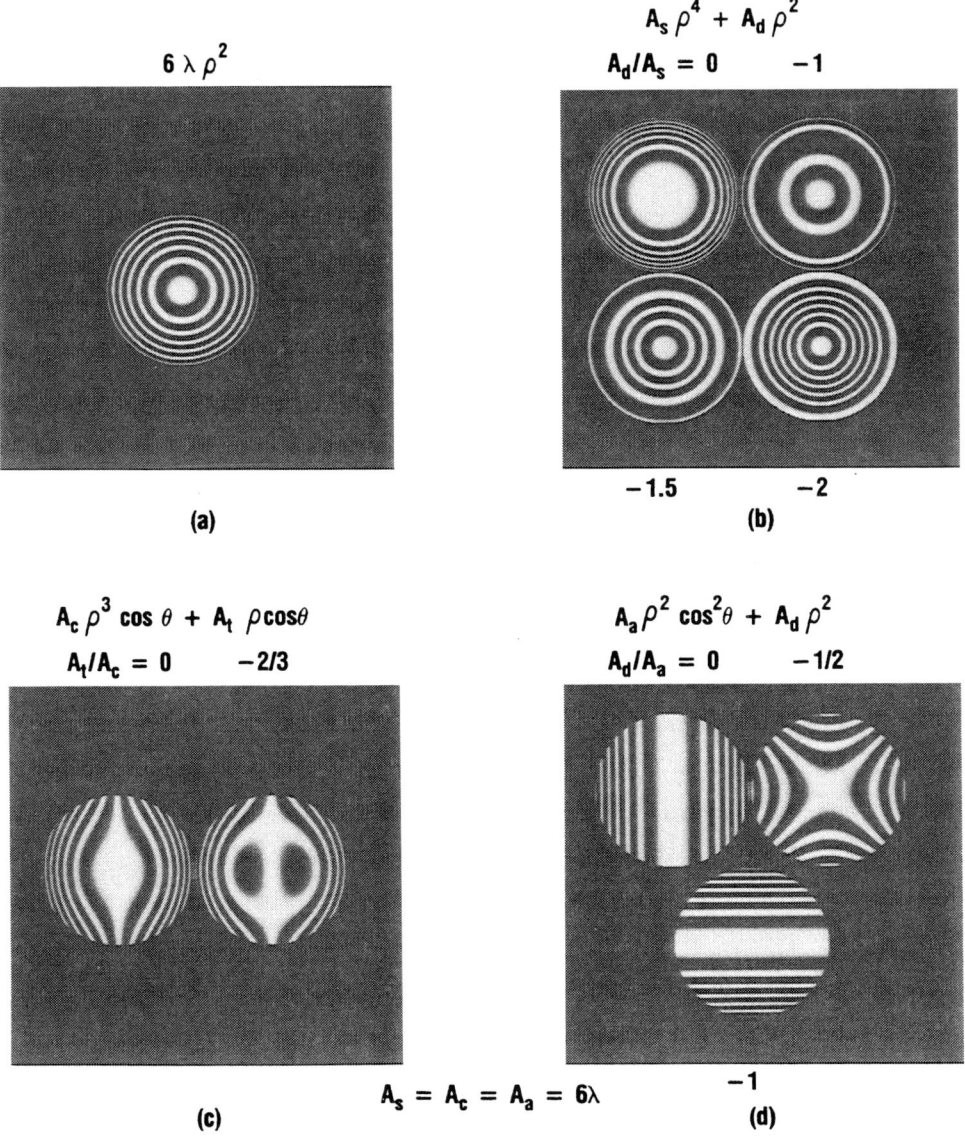

Figure 12-3. Interferograms of primary aberrations: (a) defocus, (b) spherical combined with defocus, (c) coma combined with tilt, (d) astigmatism combined with defocus. The aberrations in the interferograms are twice their corresponding values in the system under test because the test beam goes through the system twice.

represents the system when the image is observed in its marginal image plane. For a system with positive spherical aberration, its marginal focus lies farther from it than its paraxial focus (see Figure 7-1). Hence, this interferogram is obtained when the points F and C are separated from each other axially, according to Eq. (8-12), by $-48 F^2$, i.e., when F lies to the left of C by $48 F^2$. The other two interferograms, $A_d = -A_s$ and $A_d = -1.5 A_s$, represent the system when the image is observed in the minimum-aberration-variance plane (or maximum Strehl ratio for small values of A_s) and the circle-of-least-confusion plane, respectively.

Figure 12-3c shows the interferograms obtained when light is incident at a certain angle from the axis of the system such that it suffers from 3λ of coma. The fringes in this case are cubic curves. The case $A_t = 0$ corresponds to two parallel interfering beams (F and C are coincident in this case). The case $A_t = -2A_c/3$ represents the system corresponding to minimum aberration variance. A tilt aberration with a peak value of A_t may be obtained by transversally displacing C from F by $(-2FA_t, 0)$ so that C lies at the diffraction focus of the comatic diffraction pattern of the system (see Section 8.3.3 for a discussion of the diffraction focus). It may also be obtained by tilting the plane mirror M_1 by an angle A_t/a, where a is the radius of the test beam [see Eqs. (1-5) and note the factors of 2 because of the reflection of the reference beam by mirror M_1 and doubling of the system aberration in the test beam].

Figure 12-3d shows the interferograms obtained when the system suffers from 3λ of astigmatism. When $A_d = 0$ or $-A_a$, representing the system with an image being observed in a plane containing one or the other astigmatic focal line, respectively, we obtain an interferogram with straight line fringes, since the aberration depends on either x or y (but not both). However, the fringe spacing is not uniform. When $A_d = -A_a/2$, the fringe pattern consists of rectangular hyperbolas. If the system under test is aberration free but the two interfering beams are tilted with respect to each other, representing a wavefront tilt error, we obtain straight line fringes that are uniformly spaced. The fringe spacing is inversely proportional to the tilt angle.

So far we have discussed interferograms of primary aberrations when only one of them is present. These interferograms are relatively simple and the aberration type may be recognized from the shape of the fringes. It should be evident that a general aberration consisting of a mixture of these aberrations and/or others will yield a much more complex interferogram. As an example of a general aberration, Figure 12-4a shows a possible aberration introduced by atmospheric turbulence as in ground–based astronomical observations. It corresponds to $D/r_0 = 10$ as discussed in Section 11.4. On the average, the standard deviation of the instantaneous aberration introduced is given by $[0.134 (D/r_0)^{5/3}]^{1/2}$ which, for $D/r_0 = 10$, is 2.494 radians or 0.397λ. The interferogram for this aberration is shown in Figure 12-4b. When 25λ of tilt are added to the aberration, the interferogram appears as in Figure 12-4c. Doubling of the aberration as in a Twyman-Green interferometer is not considered in Figure 12-4.

References

1.   D. Malacara, ed., *Optical Shop Testing*, Wiley, New York (1977).

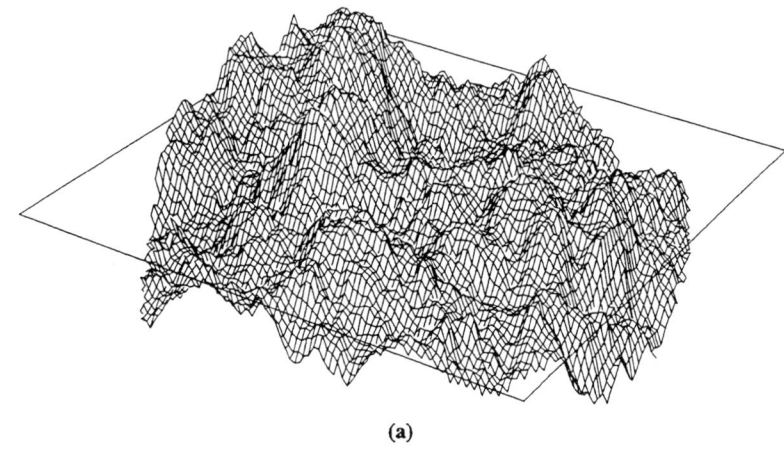

(a)

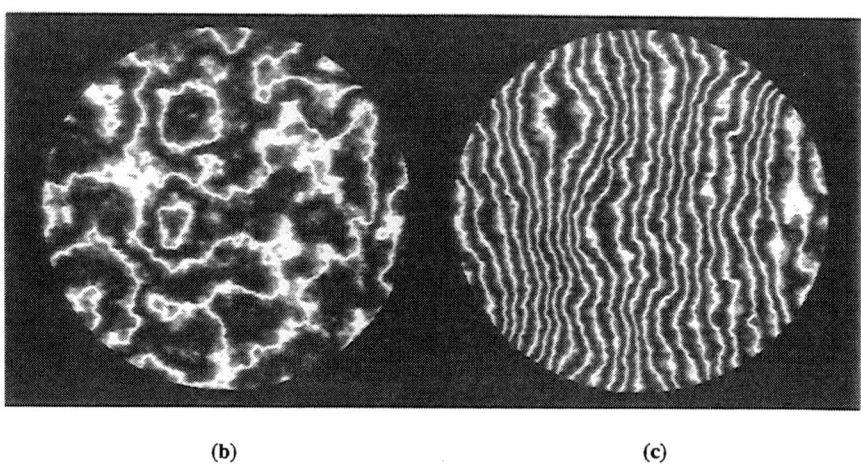

(b)   (c)

Figure 12-4. Aberration introduced by atmospheric turbulence corresponding to $D/r_0 = 10$. (a) Aberration shape. (b) Aberration interferogram. The standard deviation of the tilt-free aberration introduced by turbulence is 0.396λ. (c) Interferogram with 25λ of tilt.

Bibliography

M. Born and E. Wolf, *Principles of Optics*, Pergamon, New York, 1985.

A. E. Conrady, *Applied Optics and Optical Design*, Parts I and II, Oxford, London, 1929 (Reprinted by Dover, New York, 1957).

H. H. Hopkins, *Wave Theory of Aberrations*, Oxford, London, 1950.

F. A. Jenkins and H. E. White, *Fundamentals of Optics*, 4th ed., McGraw-Hill, New York, 1976.

M. V. Klein and T. E. Furtak, *Optics*, 2nd ed., Wiley, New York, 1986.

E. H. Linfoot, *Recent Advances in Optics*, Clarendon, Oxford, 1955.

L. C. Martin and W. T. Welford, *Technical Optics*, Vol. I, 2nd ed., Pitman, London, 1966.

E. L. O'Neill, *Introduction to Statistical Optics*, Addison-Wesley, Reading, Massachusetts, 1963.

D. J. Schroeder, *Astronomical Optics*, Academic Press, New York, 1987.

W. J. Smith, *Modern Optical Engineering*, McGraw-Hill, New York, 1966.

W. T. Welford, *Aberrations of the Symmetrical Optical System*, Academic Press, New York, 1974.

Additional References

1. V. D. Andreeva, "Fundamental aberration coefficients of an optical system, " Sov. J. Opt. Tech. **42**, 321–323 (1975).

2. T. Asakura, "Axial intensity distribution for an annular aperture with primary spherical aberration," Oyo Buturi **31**, 243–244 (1962).

3. T. Asakura and K. Fukui, "Study on the best focus with small amounts of primary spherical aberration," Oyo Buturi **31**, 221–232 (1962).

4. T. Asakura and H. Mishina, "Irradiance distribution in the diffraction patterns of an annular aperture with spherical aberration and coma," Japan. J. Appl. Phys. **7**, 751–758 (1968).

5. T. Asakura and H. Mishina, "Three dimensional distribution in the diffraction patterns of annular apertures with primary spherical aberration," Oyo Buturi **37**, 805–809 (1968).

6. T. Asakura and R. Barakat, "Annular and annulus apertures with spherical aberration and defocusing," Oyo Buturi **30**, 728–735 (1961).

7. R. Barakat, "Total illumination in a diffraction image containing spherical aberration," J. Opt. Soc. Am. **51**, 152–157 (1961).

8. R. Barakat, "The intensity distribution and total illumination of aberration–free diffraction images," *Progress in Optics*, Vol. 1, ed. E. Wolf (North Holland, 1961) pp. 67–108.

9. R. Barakat, "Rayleigh wavefront criterion," J. Opt. Soc. Am. **55**, 572–573 (1965).

10. R. Barakat, "The calculation of integrals encountered in optical diffraction theory," in *The Computer in Optical Research*, ed. B. R. Frieden (Springer, 1980) pp. 35–80.

11. R. Barakat and A. Houston, "Diffraction effects of coma," J. Opt. Soc. Am. **54**, 1084–1088 (1964).

12. R. Barakat and A. Houston, "The aberration of non–rotationally symmetric systems and their diffraction effects," Optica Acta **13**, 1–30 (1966).

13. R. Barakat and L. Riseberg, "Diffraction theory of the aberrations of a slit aperture," J. Opt. Soc. Am. **55**, 878–881 (1965).

14. L. Beiser, "Perspective rendering of the field intensity diffracted at a circular aperture," Appl. Opt. **5**, 869–870 (1966).

15. U. Bose, "Optical tolerances I. Rayleigh $\lambda/4$ criterion," Bull. Opt. Soc. India **2**, 47–51 (1968).

16. A. D. Budgor, "Exact solutions in the scalar diffraction theory of aberrations," Appl. Opt. **19**, 1597–1600 (1980).

17. S. N. Bezdidko, "The use of Zernike polynomials in optics," Sov. J. Opt. Tech. **41**, 425–429 (1974).

18. S. N. Bezdidko, "Calculation of the Strehl coefficient and determination of best focus plane in polychromatic light," Opt. Tech. **42**, 514–516 (1975).

19. S. N. Bezdidko, "Determination of the Zernike polynomial expansion coefficients of the wave aberration," Sov. J. Opt. Tech. **42**, 426–427 (1975).

20. S. N. Bezdidko, "Numerical method of calculating the Strehl coefficients using Zernike polynomials," Sov. J. Opt. Tech. **43**, 222–225 (1976).

21. S. N. Bezdidko, "Use of orthogonal polynomials in the case of optical systems with annular pupils," Opt. Spectrosc. **43**, 200–203 (1977).

22. J. Campos and M. J. Yzuel, "Axial and extra-axial responses in aberrated optical systems with apodizers, maximization of the Strehl ratio," J. Mod. Opt. **36**, 733–749 (1989).

23. H. S. Coleman and S. W. Harding, "The loss of resolving power caused by primary astigmatism, coma and spherical aberration," J. Opt. Soc. Am. **38**, 217–221 (1948).

24. G. Conforti, "Zernike aberration coefficients from Seidel and higher–order power–series coefficients," Opt. Lett. **8**, 407–408 (1983).

25. J. C. Dainty, "The image of a point for an aberration–free lens with a circular pupil," Opt. Commun. **1**, 176–178 (1968).

26. H. S. Dhadwal and J. Hantgan, "Generalized point spread function for a diffraction–limited aberration–free imaging system under polychromatic illumination," Opt. Eng. **28**, 1237–1240 (1989).

27. J. B. DeVelis, "Comparison of methods for image evaluation," J. Opt. Soc. Am. **55**, 165–174 (1965).

28. F. A. Dixon, "The influence of aberration and detector response on optical images," Proc. Phys. Soc. **75**, 713–728 (1960).

29. D. B. Fenneman and D. R. Cruise, "Far–field illumination of targets by annular apertures," J. Opt. Soc. Am. **65**, 1300–1305 (1975).

30. I. L. Goldberg and A. W. McCulloch, "Annular aperture diffracted energy distribution for an extended source," Appl. Opt. **8**, 1451–1458 (1969).

31. A. Greeve and G. C. Hunt, "The Strehl number of degraded diffraction patterns," Optik **40**, 18–23 (1974).

32. D. S. Grey, "Computed aberrations of spherical Schwarzchild reflecting microscore objectives," J. Opt. Soc. Am. **41**, 183–192 (1951).

33. A. K. Gupta, R. N. Singh and K. Singh, "Diffraction images of extended circular targets in presence of coma," Can. J. Phys. **55**, 1025–1032 (1977).

34. H. H. Hopkins and M. J. Yzuel, "The computation of diffraction patterns in the presence of aberrations," Optica Acta **17**, 157–182 (1970).

35. C. B. Hogge, R. R. Butts, and M. Burlakoff, "Characteristics of phase aberrated nondiffraction–limited laser beams," Appl. Opt. **13**, 1065–1070 (1974).

36. R. Herloski, "Strehl ratio for untruncated aberrated Gaussian beams," J. Opt. Soc. Am. **A2**, 1027–1030 (1985).

37. R. B. Johnson, "Polynomial ray aberrations computed in various lens designs programs," Appl. Opt. **12**, 2079–2082 (1973).

38. C. J. Kim and R. R. Shannon, "Catalog of Zernike polynomials," *Applied Opitcs and Optical Engineering*, eds., R. R. Shannon and J. C. Wyant, Vol. 10 pp. 193–221 (Academic Press, 1987)

39. W. B. King, "A direct approach to the evaluation of the variance of the wave aberration," Appl. Opt. **7**, 489–494 (1968).

40. W. B. King, "Dependence of the Strehl ratio on the magnitude of the variance of the wave aberration," J. Opt. Soc. Am. **58**, 655–661 (1968).

41. W. B. King and J. Kitchen, "The evaluation of the variance of the wave–aberration difference function," Appl. Opt **7**, 1193–1197 (1968).

42. W. S. Kovach, "Energy distribution in the PSF for an arbitrary pass band," Appl. Opt. **13**, 1769–1771 (1974).

43. Y. Li, "Dependence of the focal shift on Fresnel number and f number," J. Opt. Soc. Am. **72**, 770–774 (1982).

44. E. H. Linfoot, "Focal tolerances and best focal setting for model photographic images with primary spherical aberration," Optica Acta **8**, 233–253 (1961).

45. E. H. Linfoot and E. Wolf, "Diffraction images in systems with an annular aperture," Proc. Phys. Soc. (London) **B66**, 145–149 (1953).

46. D. D. Lowenthal, "Maréchal intensity criteria modified for Gaussian beams," Appl. Opt. 2126–2133 (1974). Errata, Appl. Opt. **13**, 2774 (1974)

47. D. D. Lowenthal, "Far–field diffraction patterns for Gaussian beams in the presence of small spherical aberrations," J. Opt. Soc. Am. **65**, 853–855 (1975).

48. A. Magiera and K. Pietraszkiewicz, "Position of the optimal reference sphere for apodized optical systems," Optik **58**, 85–91 (1981).

49. M. N. Malakhow, "Effect of optical system aberrations on the location of the energy center of gravity of the image of a point source," Sov. J. Opt. Tech. **45**, 131–132 (1978).

50. J. T. McCrickerd, "Coherent processing and depth of focus of annular aperture imagery," Appl. Opt. **10**, 2226–2230 (1971).

51. T. S. McKechnie, "The effect of defocus on the resolution of two points," Optica Acta **20**, 253–262 (1973).

52. J. Meiron, "The use of merit functions based on wavefront aberrations in automatic lens design," Appl. Opt. **7**, 667–672 (1968).

53. J. P. Mills and B. J. Thompson, "Effect of aberrations and apodization on the performance of coherent optical systems. I. The amplitude impulse response," J. Opt. Soc. Am. **A3**, 694–703 (1986).

54. J. P. Mills and B. J. Thompson, "Effect of aberrations and apodization on the performance of coherent optical systems. II. Imaging," J. Opt. Soc. Am. **A3**, 704–716 (1986).

55. K. Miyamoto, "Wave optics and geometrical optics in optical design," *Progress in Optics* Vol. 1, ed. E. Wolf (North Holland, 1960) pp. 31–66.

56. B. R. A. Nijboer, "The diffraction theory of optical aberrations. Part I: General discussion of the geometrical aberrations," Physica **10**, 679–692 (1943).

57. B. R. A. Nijboer, "The diffraction theory of optical aberrations. Part II: Diffraction pattern in the presence of small aberrations," Physica **13**, 605–620 (1947).

58. K. Nienhuis and B. R. A. Nijboer, "The diffraction theory of aberrations. Part III: General formulae for small aberrations: experimental verification of the theoretical results," Physica **14**, 590–603 (1949).

59. H. Osterberg and L. W. Smith, "Defocusing images to increase resolution," Science **134**, 1193–1196 (1961).

60. K. Pietraszkiewicz, "Determination of the optimal reference sphere," J. Opt. Soc. Am **69**, 1045–1046 (1979).

61. J. L. Rayces, "Exact relation between wave aberration and ray aberration," Optica Acta **11**, 85–88 (1964).

62. M. Rimmer, "Analysis of perturbed lens systems," Appl. Opt. **9**, 533–537 (1970).

63. C. J. R. Sheppard and T. Wilson, "Imaging properties of annular lenses," Appl. Opt. **18**, 3764–3768 (1979).

64. R. M. Sillito, "Diffraction of uniform and Gaussian Beams: an application of Zernike polynomials," Optik **48**, 271–277 (1977).

65. J. J. Stamnes, H. Heier, and S. Ljunggren, "Encircled energy for systems with centrally obscured circular pupils," Appl. Opt. **21**, 1628–1633 (1982).

66. W. H. Steel, "Etude des effets de aberrations et d'une obturation centrale de la pupille sur le contraste des images optiques," Rev. d' Opt. **32**, 4–26, 143–178, 269–304 (1953).

67. W. H. Steel, "The defocused image of sinusoidal gratings," Opt. Acta **3**, 65–74 (1956).

68. R. E. Stephens and L. E. Sutton, "Diffraction image of a point in the focal plane and several out–of–focus planes," J. Opt. Soc. Am. **58**, 1001–1002 (1968).

69. G. C. Steward, *The Symmetrical Optical System* (Cambridge, U. P. 1928).

70. P. A. Stokseth, "Properties of a defocused optical system," J. Opt. Soc. Am. **59**, 1314–1321 (1969).

71. S. Szapiel, "Maréchal intensity criteria modified for circular apertures with non-uniform intensity transmission: Dini series approach," Opt. Lett. **2**, 124–126 (1978).

72. S. Szapiel, "Aberration balancing techniques for radially symmetric amplitude distributions: a generalization of the Maréchal approach," J. Opt. Soc. Am. **72**, 947–957 (1982).

73. S. Szapiel, "Aberration-variance-based formula for calculating point-spread functions: rotationally symmetric aberrations," Appl. Opt. **25**, 244–251 (1986).

74. B. Tatian, "Aberration balancing in rotationally symmetric lenses," J. Opt. Soc. Am. **64**, 1083–1091 (1974).

75. C. A. Tayler and B. J. Thompson, "Attempt to investigate experimentally the intensity distribution near the focus in the error–free diffraction patterns of circular and annular apertures," J. Opt. Soc. Am. **48**, 844–850 (1958).

76. R. Tyson, "Conversion of Zernike aberration coefficients to Seidel and higher-order power–series aberration coefficients," Opt. Lett. **7**, 262–264 (1982).

77. J. J. H. Wang, "Tolerance condition for aberrations," J. Opt. Soc. Am. **62**, 598–599 (1972).

78. J. Y. Wang and D. E. Silva, "Wavefront interpretation with Zernike polynomials," Appl. Opt. **19**, 1510–1518 (1980).

79. W. T. Welford, "On the limiting sensitivity of the star test for optical instruments," J. Opt. Soc. Am. **50**, 21–23 (1959).

80. W. T. Welford, "Use of annular apertures to increase focal depth," J. Opt. Soc. Am. **50**, 749–753 (1960).

81. W. T. Welford, "Aberration tolerances for spectrum line images," Optica Acta **10**, 121–127 (1963).

82. W. B. Wetherell, "The calculation of image quality," in *Applied Optics and Optical Engineering*, eds. R. R. Shannon and J. C. Wyant (Academic Press, 1980), pp. 171–315.

83. E. Wolf, "The diffraction theory of aberrations," Rep. Prog. Phys. **14**, 95–120 (1951).

84. E. Wolf, "Light distribution near focus in an error–free diffraction image," Proc. Roy. Soc. **A204**, 533–548 (1951).

85. A. T. Young, "Photometric error analysis X. Encircled energy (total illuminance) calculations for annular apertures," Appl. Opt. **9**, 1874–1878 (1970).

86. M. J. Yzuel and F. J. Arlegui, "A study on the computation accuracy in the aberrational diffraction images," Optica Acta **27**, 549–562 (1980).

87. M. J. Yzuel and F. Calvo, "Point–spread function calculation for optical systems with residual aberrations and a non–uniform transmission pupil," Optica Acta **30**, 233–242 (1983).

Index

About the Author

Virendra N. Mahajan was born in Vihari, Pakistan, and educated in India and the United States. He received his Ph.D. degree in Optical Sciences from the Optical Sciences Center, University of Arizona, in 1974. He spent nine years at the Charles Stark Draper Laboratory in Cambridge, Massachusetts, where he started and headed an optics group. Since 1983, he has been at The Aerospace Corporation in El Segundo, California, where he works on space defense programs. He is also an adjunct professor at the University of Southern California in the electrical engineering-electrophysics department, where he teaches an advanced course in geometrical and physical optics. He was a visiting professor at the Indian Institute of Technology, New Delhi, under a grant from the United Nations Development Program during spring 1990. He has taught short courses on aberration theory at the National Central University, Chung Li, Taiwan, and at the annual meetings of the Optical Society of America and SPIE. He has published numerous papers on diffraction, aberration theory, adaptive optics, and acousto-optics. He is a member of the Optical Society of America and past chairman of its Astronomical, Aeronautical, and Space Optics technical group. He is also a member of the SPIE and a past member of its Education Committee.